Image and Signal Processing for Networked E-Health Applications

Image and Signal Processing for Networked E-Health Applications
Ilias G. Maglogiannis, Kostas Karpouzis and Manolis Wallace

ISBN:978-3-031-00481-0   paper   Maglogiannis/Karpouzis/Wallace
ISBN:978-3-031-01609-7   ebook   Maglogiannis/Karpouzis/Wallace

DOI 10.1007/978-3-031-01609-7

A Publication in the Springer series
SYNTHESIS LECTURES ON BIOMEDICAL ENGINEERING
Lecture #2

First Edition
10 9 8 7 6 5 4 3 2 1

Image and Signal Processing for Networked E-Health Applications

Ilias G. Maglogiannis
Department of Information and Communication System Engineering
University of the Aegean

Kostas Karpouzis
Institute of Communication and Computer Systems (ICCS)
University of Piraeus

Manolis Wallace
Department of Computer Science
University of Indianapolis, Athens Campus (UIA)

SYNTHESIS LECTURES ON BIOMEDICAL ENGINEERING #2

ABSTRACT

E-health is closely related with networks and telecommunications when dealing with applications of collecting or transferring medical data from distant locations for performing remote medical collaborations and diagnosis. In this book we provide an overview of the fields of image and signal processing for networked and distributed e-health applications and their supporting technologies. The book is structured in 10 chapters, starting the discussion from the lower end, that of acquisition and processing of biosignals and medical images and ending in complex virtual reality systems and techniques providing more intuitive interaction in a networked medical environment. The book also discusses networked clinical decision support systems and corresponding medical standards, WWW-based applications, medical collaborative platforms, wireless networking, and the concepts of ambient intelligence and pervasive computing in electronic healthcare systems.

KEYWORDS

Telemedicine, Biosignals, Clinical Support Systems, Coding, Standards, Web technologies, Medical Collaboration, Mobile health, Ambient Intelligence, Pervasive Computing, Virtual Reality

Contents

Preface

In all societies and in all eras, health has been an issue of major importance. Our times are, of course, no different in this, except for one thing: the kind and amount of resources available to us are augmented. During the last couple of centuries, technology has certainly been the most important new resource made available to mankind. Thus, we have witnessed the generation of a large series of new tools and devices that have helped the field of medicine progress and enhance the services it is able to offer.

In our time, it is computers and communication technology that have this role in all aspects of human activities, including medicine. As the computer-based patient record system expands to support more clinical activities, healthcare organizations are asking physicians and nurses to interact increasingly with computer systems to perform their duties. We have already seen the development of a number of medical information systems and, a few years back, we witnessed the birth of telemedicine with the utilization of networking in the field of medicine. As work in this direction progresses, medicine, medical information systems, and networking become more closely integrated, forming the area of e-Health and more specifically the field of networked and distributed e-Health.

Although e-Health has already provided a number of useful tools and applications, clearly showing its potential, it is still a field in its very early stages. As is often the case when a field has not yet matured, its literature is distributed in a wide range of publications of related fields. This makes it hard for researchers to monitor the state of the art and for new researchers to enter the field, both of which have the negative effect of slowing down the growth of the field.

In this book, we review the field of networked and distributed e-Health, hoping to provide a solid starting point for anyone interested in this marvelous science. Readers will find the chapters to be short and concise, briefly covering main achievements and estimated future directions in each important component of networked e-health applications, while at the same time offering a considerable number of current references for further reading.

We hope that the material included in this book will assist the potential reader to learn about the state of the art and future trends in the field of networked e-health applications and telemedicine and its supporting technologies.

The authors would also like to thank the publisher Joel Claypool for his endless patience and support, as well as the series editor, John Enderle, for his trust. Ilias Maglogiannis dedicates this book to Vicky, Kostas Karpouzis dedicates this book to Eleni and their two newborn daughters, and Manolis Wallace dedicates this book to Coula, Lambrini, and Nicoletta.

Ilias Maglogiannis,
Kostas Karpouzis,
Manolis Wallace
December 2005, Athens

CHAPTER 1

Introduction

For years, research in medicine was related closely only to natural sciences, as for example with chemistry in the research for new drugs. In the last decades we have seen a closer link to engineering as well, with mechanical parts being developed to substitute damaged or missing human parts, with microdevices being used as implants, with miniature cameras being utilized to avoid pervasive diagnostic examinations, and so on. With the rapid growth of networking technology and the enhancement of the communications capabilities it offers, it was only a matter of time before interest was also turned to the benefits that may be obtained from the integration of networking capabilities in medical procedures. Thus, the field of Telemedicine was born.

With computers continuing to further their penetration into everyday life, and information systems being utilized in practically every area where organizational tasks are relevant, it was only a matter of time before it became obvious that the way to go was to combine networked health applications with existing and emerging medical information systems. The new term coined just before the turn of the century in order to represent this new hybrid type of systems is e-Health.

E-Health applications quickly showed its potential, facilitating exchange of information between clinicians or between institutions, reducing costs, extending the scope and reach of medical facilities, enhancing the quality of service offered on- and off-site, providing for new means of medical supervision and preemptive medicine, and so on. Currently, the integration of medical networking and medical information systems is treated as an obvious and irrefragable rule; thus, standalone medical networking environments are no longer a reality and term telemedicine is in practice used interchangeably with e-Health.

In this book we provide an overview of the field of Networked e-Health applications and Telemedicine and its supporting technologies. We start from what is typically referred to as the lower end—that of signals. Thus, in Chapter 2 we review basic related concepts from the fields of signal processing. We focus more on the case of the electroencephalogram (EEG) and the electrocardiogram (ECG or EKG) and also look at medical signal processing and classification from the point of view of urgent medical support, where not every possible type of medical equipment is readily available. In addition to this medical imaging is discussed: building upon and extending the general purpose techniques developed in the parent field of image processing, medical image processing has provided a wealth of algorithms for the enhancement, analysis, and segmentation of medical images in particular as well as for the registration of different images to each other; we review all these issues in this chapter.

Chapter 3 brings us closer to the pure field of Telemedicine, discussing encoding for transmission of medical data. With compression being of central importance, the loss of information, as well as ways to minimize it, is discussed. Special attention is given to the utilization of wavelets, using which compression is also combined with the extremely useful and wanted capability of multiresolution presentation, which provides for great levels of flexibility in the design and application of medical communications.

Chapter 4, probably the lengthiest of the book, concludes the signal and data processing section of the book and discusses clinical decision support systems. We review their role and structure, as well as the most common approaches from the field of data mining that are used in order to automatically process medical data and provide support for medical decisions. We also discuss which types of e-Health applications each approach is most suitable for, given its characteristics.

In the last section of the book we move the discussion to topics related to the design, implementation, and operation of real life e-Health systems. Starting with Chapter 5 we review the standards that have been developed for the representation of medical data, so that their exchange between different systems may be facilitated.

Chapter 6 explains how the existing infrastructure of the Internet and the World Wide Web is exploited in the scope of e-Health, describing the way different general-purpose protocols are utilized in a purely medical scope. The important issue of data security is also discussed.

With Chapter 7 we focus specifically on collaborative platforms. These allow for clinicians to utilize the Internet in order to share files, medical data, and information in general, as well as to communicate with each other in an intuitive manner, even when they are geographically separated, thus facilitating their cooperation.

Wireless and *ad hoc* networks have received a lot of attention lately and are expected to have a greater role in everyday life in the future. In Chapter 8 we review their role in telemedicine together with their supporting protocols and standards.

In Chapter 9 we present the concepts of ambient intelligence and pervasive computing in electronic healthcare systems. The concepts refer to the availability of software applications and medical information anywhere and anytime, and the invisibility of computing modules when they are hidden in multimedia information appliances, which are used in everyday life.

Chapter 10 concludes the book with a reference to virtual reality systems and techniques and their application toward more intuitive interaction in a networked medical environment.

Overall, we believe we have provided a complete and balanced view of the field; we hope our readers will think so too, and that they find this book to be a good point to start and a useful reference to come back to.

CHAPTER 2

Signal Processing for Telemedicine Applications

2.1 INTRODUCTION TO SIGNAL PROCESSING

A very general definition of a signal is a "measurable indication or representation of an actual phenomenon," which in the field of medical signals refers to observable facts or stimuli of biological systems or life forms. In order to extract and document the meaning or the cause of a signal, a medical practitioner may utilize simple examination procedures, such as measuring the temperature of a human body or have to resort to highly specialized and sometimes intrusive equipment, such as an endoscope. Following signal acquisition practitioners go on to a second step, that of interpreting its meaning, usually after some kind of signal enhancement or "pre-processing" that separates the captured information from noise and prepares it for specialized processing, classification, and recognition algorithms. It is only then that the result of the acquisition process reveals the true meaning of the physical phenomenon that produced the signal under investigation.

As a general rule, the particular techniques used in all the above mentioned steps depend on the actual nature of the signal and the information it may convey. Signals in medicine-related applications are found in many forms, such as bioelectric signals that are usually generated by nerve and muscle cells. Besides these, tissue bioimpedance signals may contain important information such as tissue composition, blood volume and distribution, or endocrine activity, while bioacoustic and biomechanical signals may be produced by movement or flow within the human body (blood flow in the heart or veins, blood pressure, joint rotation or displacement, etc.). While acquiring these kinds of signals is usually performed with unintrusive examinations, biomagnetic signals, produced at the brain, heart, and lungs, and biochemical signals can be collected only with specialized instruments and in controlled hospital or research center environments.

While medical signal processing techniques can sometimes be performed on raw, analog signals, advanced, frequency-domain methods typically require a digitization step to convert the signal into a digital form. Besides catering for the deployment of techniques that would be otherwise impossible, digital signals are much more efficient when it comes to storing and transmitting them over networks or utilizing automated feature-extraction and recognition techniques. This process begins with acquiring the raw signal in its analog form, which is then fed into an analog-to-digital (A/D) converter. Since computers cannot handle or store continuous data, the first step of the conversion procedure is to produce a discrete time series from the analog form of the raw signal. This step is known as "sampling" and is meant to create a sequence of values sampled from the original analog signals at predefined intervals, which can faithfully reconstruct the initial signal waveform. In order for this to happen, the sampling frequency must be at least double the signal bandwidth; this requirement is known as the Shannon theorem and, in theory, it is the deciding requirement for the sampling process to produce a faithful representation of the captured signal. In practice, however, picking the sampling frequency with only Shannon's theorem in mind may lead to other equally important problems such as aliasing or noise replication.

The second step of the digitization process is quantization, which works on the temporally sampled values of the initial signal and produces a signal that is both temporally and quantitatively discrete. This means that the initial values are converted and encoded according to properties such as bit allocation and value range. Essentially, quantization maps the sampled signal into a range of values that is both compact and efficient for algorithms to work with. In the case of capturing a digital image or scanning a printed one the process of sampling produces a set of values, each of which represents the magnitude of visible light falling on each single sensor cell integrated over the area of the sensor. Depending on the number of bits allocated for the representation of the digital image, each single value is then quantized over the relevant range of values; for example, if 8 bits per pixel are available, 256 discrete levels of luminosity can be represented and, thus, the sampled value is quantized to fit into the interval [0, 255].

2.2   SIGNAL PROCESSING APPLICATIONS

Although there exist a number of algorithms that operate on the sampled and quantized digital signals as is, most of the techniques that extract temporal information from and recognize patterns in a signal are performed in the frequency domain. The foundation of these algorithms is the fact that any analog or digital signal can be represented as an integral or sum of fundamental sine functions with varying amplitudes and phases. The reversible transformation between the initial representation and that of the frequency domain is given by the Fourier transform (FT) [10], which is defined in both continuous and discrete forms. An interesting alternative to the frequency representation is to apply the FT to the correlation function, resulting in the power spectral density function (PSD) [17], which finds a prominent use in recognizing pathological states in electroencephalograms (EEGs).

2.2.1   Electrocardiography

The electrocardiogram (ECG) signal represents the electrical activity produced by the heart and acquired at skin surface level and has been available from nearly the beginning of the 20th century [1]. Actually some of the techniques used to capture the signal (usually at the arms and legs) and to label salient features originate from the pioneering work of Einthoven [7]; ECGs were also among the earliest acquired signals to be transferred to remote locations over phone lines. In recent capture stations dedicated hardware is used to perform the A/D conversion of the differential signals captured at as many as 12 sites of the human body. These 12 "leads" are then combined to produce the conclusive measurements, which are in the following fed to classification algorithms. ECGs are typically used to determine the rate and rhythm for the atria and ventricles and pinpoint problems between or within the heart chambers. In addition to this, this signal can illustrate evidence of dysfunction of myocardial blood perfusion, or ischemia, or chronic alteration of the mechanical structure of the heart, such as enlarged chamber size [19].

Emergency ECG processing caters for early detection of possibly severe pathologies, since the permanent or temporary effects of even minor cardiac dysfunctions on the waveform can be detected and classified. Regarding acquisition, bulky Holter [9] devices built in the 1960s are currently substituted with compact devices that cater for easy deployment at mobile sites. Initially, Holter probes were combined with 24-h recording and playback equipment in order to help identify periods of abnormal heart rate, possibly attributed to a developing heart block. These devices possessed minimal heart beat segmentation capabilities and could also illustrate results visually and audibly; in addition to this, identification of premature ventricular complexes (PVCs) and relevant treatment is also possible and is recently enhanced by high-speed, hardware-assisted playback. Despite the decline in its use, emergency ECG is still a very useful early detection tool, based on devices with dedicated memory and storing capabilities, capable to provide automated signal processing and classification without the need for a host computer. High-resolution ECGs are the most recent signal acquisition and processing improvement tools [2], capable to record signals of relatively low magnitude that occur after the QRS complex (the deflection in the ECG that represents ventricular activity of the heart) but are not evident on the standard ECG and can be related to abnormally rapid heart rate (ventricular tachycardia) [20].

ECG signal processing lends well to knowledge-based techniques for classification. As a result the best known and most widely used such methods have been employed in this framework, including Bayesian [26] algorithms and Markov models [4]. Neural networks (NNs) [6, 21] have also been put to use here with often better results, since NNs can be designed to operate with and handle incomplete data, which is quite often the case with ECG data sets.

2.2.2   Electroencephalography

Electroencephalograms capture the very low electrical activity produced by the brain [3, 8, 16]. Initially, EEGs were used to detect epilepsy-related conditions, but in the process, their use was extended to nonpathological, behavioral situations, such as detecting and studying the ground truth of emotion-related stimuli in the human brain [23]. Another factor of improvement in the study of EEGs is related to the means of analysis, which developed from being merely visual and subjective to semiautomated techniques working on signal waveform properties such as amplitude and phase. Acquisition is performed using recording systems consisting of conductive electrodes, A/D converters, and an attached recording device. Classification is one of the major signal processing tasks performed on the EEG signal: usually this is carried out by segmenting the signal into "chunks," and then calculating the closest "bin" in which the particular chunk is classified. Statistical analysis follows, and the distribution of chunks along the different classes is checked against predefined patterns. This last stage of control is not performed on the distribution curve *per se*, but more often on its quantitative control parameters, such as standard deviation or moments.

Regarding frequency domain signal processing, the most usual algorithms put to use are the fast Fourier transform (FFT) and analysis of the PSD. Both techniques are best suited for the extraction of specific features, but perform badly and essentially destroy some of the signal parameters. For example, FFT is typically computed over a predefined, *albeit* short time window, and thereby does not take into account variation within that time frame; nevertheless, it provides a reliable estimation of the PSD, which in turn is related to changing psychological states of the subject. In order to overcome the windowing effects, one may opt to move to autoregressive models.

2.3   MEDICAL IMAGE PROCESSING

Medical image processing can be thought of as a specialization of signal processing techniques and their application to the two-dimensional (2D) domain, in the case of still images, or three-dimensional (3D) domain for volumetric images or video sequences. A wide number of issues exist here such as image acquisition and coding, image reconstruction (e.g., from slices), image enhancement and denoising, and image analysis. On a higher level, automatic or user-assisted techniques may be used to segment and detect specific objects in an image, e.g., organs in a computed axial tomography (CAT) slice or process and visualize volumetric data [5].

Progress in medical imaging coincides not only with the advent of maturing processing techniques, but also with the introduction of better acquisition equipment and the integration of knowledge-based strategies. Regarding complexity and tackling of novel concepts, the arrival of magnetic resonance imaging (MRI) and other related acquisition techniques that produce

3D or higher resolution images paved the way for visualization applications and model-driven approaches.

In the field of 2D imaging image registration, segmentation, and pattern recognition and classification are still in the forefront of interest. In the case of segmentation, knowledge-based approaches [25] that are constantly integrated even into specialized equipment cater for tackling problems such as edge detection and region growing [12], and subsequent object boundary detection. CAT and MRI scans are eminent examples of such techniques, since lower level segmentation of image intensity can be combined with knowledge of tissue structure and positioning to cater for organ detection and segmentation. In addition to this, deformable models [13, 14] can be employed to track the deformation of organs such as the heart and model several functional organ parameters, while successive scan slices can be used to reconstruct the 3D volume of an internal organ [22]. It has to be noted here that the initial thickness of these slices would not cater for faithful, usable reconstruction; however, modern equipment can produce multiple successive slices in the order of a few millimeters, hence only minute volume details are lost.

Image texture is another useful tool, especially in cases where tissue density is mapped to pixel intensity (e.g., ultrasound [15]), but can also be used for motion analysis (e.g., in echocardiograms). Several issues still exist here, such as correct calibration of the instrument and provision of contrast enhancement tools that cater for faithful mapping of light intensity to tissue density; result of inconsistencies between different acquisition devices and setups result in texture information being put to better use in segmentation problems than in, e.g., modeling organ tissue using knowledge-based techniques.

A very important concept of medical imaging arises from the fact that information related to the same organ or tissue can be collected from a multitude of sources; as a result, there is the need to productively integrate these inputs, while still retaining the interrelation of what is provided from each source. With the arrival of multiresolution data, the problem of image matching and registration [11] became more apparent, since there could be cases where the same observation could be contained in both 2D slices and 3D volumetric scans, taken from different points of view, during different examinations and with different resolution. One approach [18], which tackled matching manually segmented surfaces from the brain, worked on CAT and MRI scans, as well as positron emission tomography (PET) data. Here, since temporal and spatial resolutions were diverse, the best solution would come from simple geometrical distance metrics in order to calculate the best matching transformation between the different sets of data [24].

REFERENCES

[1] J. Bailey, A. Berson, and A. Garson , "Recommendations for standardization and specifications in automated electrocardiography: bandwidth and digital signal processing. A report for health professionals by an ad hoc writing group of the committee on

electrocardiography and cardiac electrophysiology of the Council on Clinical Cardiology, American Heart Association," *Circulation*, vol. 81, p. 730, 1990.

[2] E. Berbari, "High-resolution electrocardiography," *CRC Crit. Rev. Bioeng.*, vol. 16, p. 67, 1988.

[3] J. Bronzino, "Quantitative analysis of the EEG: general concepts and animal studies," *IEEE Trans. Biomed. Eng.*, vol. 31, no. 12, p. 850, 1984.

[4] A. Coast, R. Stern, G. Cano, and S. Briller, "An approach to cardiac arrhythmia analysis using hidden Markov models," *IEEE Trans. Biomed. Eng.*, vol. 37, pp. 826–835, 1990.doi:10.1109/10.58593

[5] J. Duncan, "Medical image analysis: progress over two decades and the challenges ahead," *IEEE Trans. PAMI*, vol. 22, no. 1, pp. 85–106, 2000.

[6] L. Edenbrandt, B. Devine, and P. Macfarlane, "Neural networks for classification of ECG ST-T segments," *J. Electrocardiol.*, vol. 25, no. 3, pp. 167–173, 1992.doi:0.1016/0022-0736(92)90001-G

[7] W. Einthoven, "Die galvanometrische Registrirung des menschlichen Elektrokardiogramms, zugleich eine Beurtheilung der Anwendung des Capillar-Elecktrometers in der Physiologie," *Pflugers Arch. Ges. Physiol.*, vol. 99, p. 472, 1903.doi:10.1007/BF01811855

[8] A. Givens and A. Remond, Eds., "Methods of analysis of brain electrical and magnetic signals," in *EEG Handbook*, vol. 1. Amsterdam: Elsevier, 1987.

[9] N. Holter, "New method for heart studies: continuous electrocardiography of active subjects over long periods is now practical," *Science*, vol. 134, p. 1214, 1961.

[10] S. Kay, *Modern Spectral Estimation: Theory and Application*. Englewood Cliffs, NJ: Prentice-Hall, 1988.

[11] J. Maintz and M. Viergever, "A survey of medical image registration," *Med. Image Anal.*, vol. 2, no. 1, pp. 1–16, 1998.doi:10.1016/S1361-8415(01)80026-8

[12] A. Martelli, "An application of heuristic search methods to edge and contour detection," *Comm. ACM*, vol. 19, pp. 73–83, 1976.doi:10.1145/359997.360004

[13] T. McInerney and D. Terzopoulos, "Deformable models in medical image analysis: a survey," *Med. Image Anal.*, vol. 1, no. 2, pp. 91–108, 1996.doi:10.1016/S1361-8415(96)80007-7

[14] D. Metaxas and D. Terzopoulos, "Constrained deformable superquadrics and nonrigid motion tracking," in *Computer Vision and Pattern Recognition*, G. Medioni and B. Horn, Eds. IEEE Computer Society Press, 1991 pp. 337–343.doi:full_text

[15] D. Morris, "An evaluation of the use of texture measures for tissue characterization of ultrasound images of in vivo human placenta," *Ultrasound Med. Biol.*, vol. 14, no. 1, pp. 387–395, 1988.doi:10.1016/0301-5629(88)90074-9

[16] E. Niedermeyer and F. Lopes da Silva, *Electroencephalography: Basic Principles, Clinical Applications and Related Fields*, 3rd ed. Philadelphia: Lippincott, Williams & Wilkins, 1993.

[17] T. Ning and J. Bronzino, "Autoregressive and bispectral analysis techniques: EEG applications," *IEEE Eng. Med. Biol. Mag.*, vol. 9, Special Issue, p. 47, 1990. doi:10.1109/51.62905

[18] C. Pelizzari, G. Chen, D. Spelbring, R. Weichselbaum, and C. Chen, "Accurate three-dimensional registration of CT, PET, and/or MR images of the brain," *J. Comput. Assist. Tomogr.*, vol. 13, pp. 20–26, 1989.

[19] T. Pryor, E. Drazen, and M. Laks, Eds., *Computer Systems for the Processing of Diagnostic Electrocardiograms*. Los Alamitos, CA: IEEE Computer Society Press, 1980.

[20] M. Simson, "Use of signals in the terminal QRS complex to identify patients with ventricular tachycardia after myocardial infarction," *Circulation*, vol. 64, p. 235, 1981.

[21] R. Silipo and C. Marchesi, "Artificial neural networks for automatic ECG analysis," *IEEE Trans. Signal. Proc.*, vol. 46, no. 5, pp. 1417–1425, 1998. doi:10.1109/78.668803

[22] L. Staib and J. Duncan, "Model-based deformable surface finding for medical images," *IEEE Trans. Med. Imag.*, vol. 78, no. 5, pp. 720–731, 1996. doi:10.1109/42.538949

[23] J. Taylor and C. Mannion, Eds., *New Developments in Neural Computing*. Bristol, England: Adam Hilger, 1989.

[24] J. P. Thirion, "New feature points based on geometric invariants for 3D image registration," *Int. J. Comput. Vis.*, vol. 18, no. 2, pp. 121–137, 1996. doi:10.1007/BF00054999

[25] J. Tsotsos, "Knowledge organization and its role in representation and interpretation for time-varying data: the ALVEN system," *Comput. Intell.*, vol. 1, no. 1, pp. 16–32, 1985.

[26] J. Willems and E. Lesaffre, "Comparison of multigroup logistic and linear discriminant ECG and VCG classification," *J. Electrocardiol.*, vol. 20, pp. 83–92, 1987. doi:10.1016/0022-0736(87)90036-7

CHAPTER 3

Medical Data Encoding
for Transmission

3.1 INTRODUCTION

Medical information, especially in time-varying or multidimensional and multiresolution forms, typically creates large databases and requires relevant storage media and strategies. High quality stereo sound, for example, is sampled in 44 kHz, with 16 bits allocated per sample and two channels combined to produce the final file; a simple multiplication shows that this acquisition scenario produces data at 1.4 Mb/s. In the same framework, uncompressed NTSC TV signal at 30 frames/s utilizes almost 300 Mb/s, while for quarter-frame sizes, e.g. webcams, this drops to almost 73 Mb/s. Different applications may impose additional requirements: for example, machine-learning techniques may well operate offline and thus alleviate the need for real time availability, while emergency processing for early detection needs as much as data as possible on demand. In the case of transmission, the choice of medium is another critical factor, since the difference in bandwidth capacity between T3 networks and normal telephony networks (PSTN) is in the order of 1000 to 1. Data compression may be an obvious option here, but this also presents a number of interesting questions, especially when lossy compression techniques are put to use, thereby removing possibly critical information from the data set. In addition to this, the huge amounts of data produced by the acquisition equipment may also be a serious obstacle [2].

Compression of digital information comes in two forms: lossy, where data is encoded reversibly and, as a result, decoding will faithfully reproduce the initial data set, and lossless, in which facets of the data set are sacrificed to achieve bigger gains in file size and bandwidth for transmission. There are apparent advantages in either category, but the most prominent of them are the trade-off between compression reversibility and complexity, on one hand and resulting size of information and required bandwidth, on the other. In the case of lossless encoding, typical compression rates are in the order of 2 to 1, meaning that the size of the file produced by the compression algorithm is half of the initial file or that the size gain is only 50%; however, all information contained in the original data set is retained and can be retrieved with minimal decoding effort. When lossy algorithms are employed, a portion of the data is deemed either unnecessary to retain or "almost" correlated with another portion and thus can be dismissed. In the case of audio compression using the MPEG-1 standard, for example, besides the other compression schemes proposed by the standard itself, the more advanced encoders exploit what is known as the "psychoacoustic phenomenon." Here the existence of particular audible frequencies in sounds is thought to prevent the human ear from receiving neighboring frequencies, which as a result can be dropped from the original sound spectrum. While this process effectively alters the sound without the possibility to reproduce faithfully, the difference between the original waveform and the one reconstructed from the encoded file is thought to be unnoticeable for practical purposes. In the case of images, the human eye is thought to be more sensitive to the brightness (luminance) value of a pixel than to the color (chrominance) information; as a result, a portion of color information can be discarded (subsampled) without introducing perceptible deterioration. Here lossless techniques can provide compression ratios in the order of 20 to 1, with minimal visible artifacts in the image; as a result, for most applications, lossless image compressions techniques with measurable and small introduction of error are preferred [5] [11].

3.2   DATA COMPRESSION

Besides the lossy/lossless division, one may categorize compression efforts with respect to their analogy to signal representation and processing, i.e., distinguish between algorithms operating on temporal representations and the frequency domain. Usually the particular technique that is used depends on the actual signal to be compressed. In the case of electrocardiograms (ECGs) a compression ratio of 7.8 has been reported [9] using differential pulse code modulation (DPCM), an algorithm that uses linear prediction in order to decorrelate the samples of the input signal. The rationale behind this is that data samples with higher probabilities contribute less to what is already known and, since they occur more often than the rest, are assigned to smaller "codewords," i.e., are represented with less bits in a look-up table. Reversely, sample values that do not arise as frequent and, hence, have small probabilities can be assigned to larger representations, a fact that does not harm file size since the values in question are infrequent. This lossless process, which is called Huffman encoding, can be performed in both 1D and 2D signals, making it a very important tool for data compression. It has to be noted though that the remaining modules of the DPCM compression algorithm, and especially quantization of the values of the error sequence that is produced by the prediction/decorrelation step, introduce irreversible loss of information that, however, permits higher compression rates.

DPCM is a typical example of algorithms used to prepare the signal for either storage or transmission; signal encoding may also be performed to segment the signal or detect prominent features from it. In this framework the amplitude zone time epoch coding (AZTEC) algorithm [3] can be used to assist in analyzing ECGs, producing QRS detection automatically.

One may describe this method as "adaptive downsampling," where temporally static or almost static information within a specific time window is replaced by a constant value. To reconstruct the waveform, linear interpolation between the remaining samples is sufficient. However, this produces a lower resolution waveform that has an inherently "jagged" look.

When frequency-domain data compression techniques are put to use, regardless of the dimensionality of the data, input signals are first segmented to blocks or "chunks" of predefined size and then each of these chunks is transformed into the frequency domain. The Discrete Cosine Transform (DCT) [1] is the algorithm most often employed, especially in the case of images and extended to videos in the sense that they are approached as a sequence of frames. A number of popular video compression standards (MPEG-1, −2, and −4, as well as H.26x [8]) utilize DCT at some stage of encoding. In the case of images, JPEG is the most usual case of lossy compression and is based on the energy-packing characteristics of DCT, performing processes where amounts of information are irreversibly lost, resulting in huge compression rates. Its modes of operation are further divided into baseline, extended progressive, and extended hierarchical. There also exists a lossless variant of JPEG that operates using predictive algorithms, but fails to produce similar compression ratios as in the lossy case. In the beginning of the JPEG algorithm, the input image is segmented into two sets of blocks of 8×8 pixels, one for the luminance and one for the chrominance channel of the image, starting from the top left of the image; these blocks are in the following transformed to produce the 2D DCT, resulting in 64 DCT coefficients. Irreversible loss of information occurs in the next step, where the DCT coefficient values are quantized and thus a part of the variation among them is lost. This step hurts the larger values of the DCT coefficients in the block in question, resulting in loss of higher frequencies, which means that edges are blurrier and object boundaries are not so well separated. The amount of rounding off in each quantization is the deciding factor for the amount of compression and the decrease of the signal-to-noise ratio, which is a metric of image quality versus any introduced noise. In the final step of the algorithm, DCT coefficients within a block are arranged in a zigzag fashion and fed into a Huffman encoder to produce the final encoded image.

In the case of video sequences, the MPEG standard exploits both spatial and temporal redundancy. When handling a single frame, the encoding scheme can be very close to JPEG compression; for subsequent frames, the standard adopts the intra- and intercoding schemes. Here an MPEG GOP (group of pictures) is made of I- (intra), P- (predictive), and B- (bidirectional) coded frames. An I-frame has the role of a "key frame," in the sense that all processing is independent of following and previous information, and is uncompressed or compressed in a lossless fashion. P- and B-frames are intercoded reference frames, where temporal redundancy on reference frames is removed by performing motion estimation: a P-frame is coded based on its preceding I- or P-frame, while a B-frame is coded using block-based information from both of the preceding and the following I- and/or P-frames. Combination of the two methods is performed in MJPEG compression, where each frame of the video sequence is compressed individually using the JPEG compression algorithm.

3.3   WAVELET COMPRESSION

Most recent trends in the field of medical image compression are focused on the wavelet transformation and compression technology [12], leading to the introduction of the new image compression standard, JPEG 2000 [10], that, however, is yet to enjoy wide adoption. Wavelets cater for a multiresolution representation of the data set [6] [7], irrespective of dimensionality and without loss of information, as is the case with DCT-based techniques. This very important characteristic enables users to browse through a general view of the complete data set, e.g., a volumetric 3D scan, and focus on a specific portion which is decoded and presented in greater detail. Scalable rendering is of utmost importance in cases of large data sets, which can be in the order of hundred of millions of volumetric pixels, since it caters for visualization in computers with everyday capabilities as well as higher level observation [12].

Signal-wise, the wavelet transform is a form of subband coding, where the transformed signal information contained in both high-pass and low-pass parts can be repeatedly divided in a similar fashion. This process can be repeated, effectively producing a hierarchical tree structure in which each segment can be separately encoded; in fact, the wavelet transform concept can be thought as a windowed-preparation technique, with variable-sized regions, that allows the use of longer intervals where more precise low-frequency information is required and shorter regions where we want high-frequency information. In addition to this, wavelets provide means for local processing of the signal, without the need to decode the complete waveform and can also easily reveal unique higher level information like discontinuities in higher derivatives or self-similarity [4].

REFERENCES

[1]   N. Ahmed, T. Natarajan, and K. Rao "Discrete cosine transform," *IEEE Trans.Comput.*, pp. 90–93, 1974.

[2]   A. Cetin and H. Koymen, "Compression of digital biomedical signals," in *The Biomedical Engineering Handbook*, J. Bronzino, Ed., 2nd ed. Boca Raton: CRC Press LLC, 2000.

[3]   J. Cox, "AZTEC: a preprocessing program for real-time ECG rhythm analysis," *IEEE Trans. Biomed. Eng.*, vol. 15, pp. 128–129, 1968.

[4]   J. Crowe, "Wavelet transform as a potential tool for ECG analysis and compression," *J. Biomed. Eng.*, vol. 14, pp. 268–272, 1992.

[5]   M. Karczewicz and M. Gabbouj, "ECG data compression by spline approximation," *Signal Process.*, vol. 59, pp. 43–59, 1997.doi:10.1016/S0165-1684(97)00037-6

[6]   S. Mallat, "A theory for multiresolution signal decomposition: the wavelet representation," *IEEE Trans. PAMI*, vol. 11, pp. 674–693, 1989.

[7]   P. Morettin, "From Fourier to wavelet analysis of time series," in *Proc. Comput. Stat.*, New York: Physica, 1996, pp. 111–122.

[8]   I. Richardson, *H.264 and MPEG-4 Video Compression: Video Coding for Next-Generation Multimedia*. England: Wiley, 2003.

[9]   U. Ruttiman and H. Pipberger, "Compression of the ECG by prediction or interpolation and entropy coding," *IEEE Trans. Biomed. Eng.*, vol. 26, no. 4, pp. 613–623, 1979.

[10] D. Taubman, and M. Marcellin, *JPEG2000: Compression Fundamentals, Standards and Practice*. Norwell, MA: Kluwer Academic, 2001.

[11] J. Willems, "Common standards for quantitative electrocardiography," *J. Med. Eng. Techn.*, vol. 9, pp. 209–217, 1985.

[12] Z. Xiong, X. Wu, S. Cheng, and J. Hua, "Lossy-to-lossless compression of medical volumetric data using 3-D integer wavelet transforms," *IEEE Trans. Med. Imag.*, vol. 22, pp. 459–470, Mar. 2003.doi:10.1109/TMI.2003.809585

CHAPTER 4

Clinical Decision Support Systems for Remote and Commuting Clinicians

4.1 INTRODUCTION

This chapter presents a special branch of expert systems called clinical decision support systems (CDSSs). In general, defining what is and what is not a CDSS is not an easy task [1]. Still, one can generally say that the term generally refers to automated systems that process medical data such as medical examinations and provide estimated diagnoses [2]. The estimations are often based on the analysis of details that elute the human eye as well as large amounts of medical history that humans cannot possibly consider. Although such systems typically do not reach 100% success, which means that they cannot substitute the clinician, the input they provide is extremely helpful as an independent source of evidence concerning the correct diagnosis [3].

CDSSs have an augmented role in the framework of networked e-Health [4]. Clinicians in small-scale facilities located at remote and isolated locations, as well as clinicians providing their services at the site of events that require medical attention before transportation can be allowed, have to face an important handicap: they do not have access to the large medical knowledge bases, such as medical libraries, warehouses of medical history, or the benefit of having other clinicians around to consult for a second opinion. In this case, the support of an automated system that can refer to such medical knowledge bases and provide a second opinion is certainly an important help.

Although a CDSS is a very complex software system, its operation in the framework of e-Health, as well as the steps typically followed in its design and development, can be generally broken down as shown in Fig. 4.1. As shown, information extracted from a patient running various medical tests—this could be from something as simple as body temperature to detailed blood analysis results or f-MRI images—is processed by the clinician in order to reach a conclusion regarding the diagnosis. In this process the clinician may also utilize the networking capabilities offered by the system to retrieve data from distributed electronic health records and/or consult other clinicians who may be located at different locations. The CDSS attempts to simulate the same process. Thus, the extracted information is processed by the

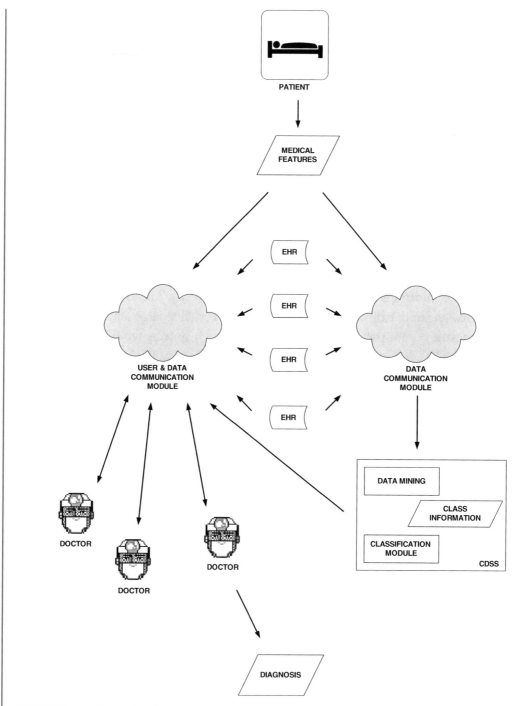

FIGURE 4.1: General architecture of a clinical decision support system

CDSS in order to reach a diagnosis. This diagnosis is provided by the classification module, which maps available information to one of the known classes of diagnoses. This information about the known classes is either provided manually by human experts or, most often, extracted automatically through processing of the medical history contained in the distributed electronic health records by a data mining module. The estimate produced by the CDSS is yet another input made available to the clinicians toward their diagnosis.

The most important processes in the development and operation of a CDSS are i) the acquisition of the information regarding the diagnosis classes and ii) the actual classification of a given case to a diagnosis. The two steps are actually closely related to each other, as the type of classifier chosen in most cases also indicates the training methodology to use. Although there is an extremely wide variety of classification methodologies that one may choose to apply, the most well known and widely applied genres of approaches can be briefly summarized and categorized as follows:

1. The examined case is compared directly to other cases in the medical history, and similarities are used in order to provide a most probable diagnosis.

2. Different types of artificial neural networks (ANNs) are trained based on available medical history, so that the underlying data patterns are automatically identified and utilized in order to provide more reliable classification of future data.

3. Information extracted automatically from medical history, or provided manually by human experts, is organized so that diagnosis estimation is provided in a dialogical manner through a series of question/answer sessions.

4. Multiple simple classifiers are combined to minimize the error margin.

5. Information provided by human experts in the form of simple rules is utilized by a fuzzy system in order to evaluate the case in hand.

Each one of these approaches has its characteristics, benefits, and drawbacks; these are discussed in detail in the following section. Especially for the case of remote or commuting clinicians, in addition to issues related to the performance of the CDSS, another important issue to consider is that of the location of the different components of the system. Depending on whether the classification information is located at a centralized system or stored on a portable device as well as depending on whether the classification system itself is portable or operates centrally and exchanges information with the clinician over a network, the types of classifiers, classification information structures, and data mining methodologies that can be applied in practice differ, based mainly on the networking resources that they require.

4.2   CLASSIFICATION SYSTEMS

Classifiers are a breed of software that has been studied for long in the field of artificial intelligence. Depending on the application, input data format, and computing resources available, a wide variety of methodologies can be selected in the literature [5, 6]. Given the diverse problems and circumstances that distributed medical applications have to deal with, many of these come in handy in different situations. In the following we review the main categories of classification systems that find application in a CDSS framework.

4.2.1 k-Nearest Neighbors

The k-nearest neighbors methodology, often referred to as k-NN, constitutes a breed of classifiers that attempt to classify the given patient data by identifying other similar cases in the database of medical history. The simplest, as well as most common, case is when all considered medical features are scalars. In that case, assuming that n features are available in total, one can consider all cases to be positioned perfectly and accurately in an n-dimensional space. Using an Euclidean distance it is possible to identify the historical medical case(s) that is closest to the one in question; the classification (diagnosis) for this medical case is then considered as the classification for the case in question as well. Parameter k is the number of similar cases to consider in order to reach a diagnosis. This simplistic approach has a number of drawbacks, the most important of which are described below.

The n-dimensional space described above is rarely partitioned perfectly in regions belonging to a single class, i.e., to regions corresponding to medical cases all diagnosed in the same way. Quite the contrary, in most regions of this space a number of classes coexist. In the quite common situation that not all cases in the considered neighborhood correspond to the same diagnosis, the diagnosis appearing the most times is the one selected by the system.

An extension to this approach is the weighted k-NN approach. This approach additionally considers the distance (or similarity) between the different cases in the neighborhood and the one in question, thus allowing most similar cases to affect the final decision to a greater extend.

Of course, in all the above we assume that the medical features measured are all scalars, and thus that it is possible to directly map the problem in the n-dimensional space. Of course, this is not always the case in medicine. Electrical signals such as electrocardiograms (ECGs) and images such as computed tomography (CT) scans are also medical features that often need to be considered but cannot be equally easily mapped to a geometrical space. Still, this does not mean that k-NN based approaches cannot be applied. In such cases, one needs to define a metric for each measured feature. In the above example, all one needs to define is a function providing the difference between two ECGs and another one providing the difference between two CT scan images. Typically, although differences are not that easy to define, it is quite simpler to define similarities. Thus, usually one identifies what it is that makes two medical signals or images similar and then defines a metric quantifying the inverse of this.

The k-NN methodology has found many applications in the field of CDSSs. For example, the following works are based on k-NN based approaches [7–10].

The applicability of k-NN and its variations in networked health applications are governed mainly by the need to compare a sample with all samples in the medical history in order to reach a conclusion. This makes it practically impossible to have the classification module located separately from the class information, as it would require the transmission of large amounts of data in every run. Therefore, it is the medical information about the current case that needs to be transmitted to a centralized system. This makes k-NN an excellent choice when these medical features are small in size, but not that practical when large signals and high resolution medical images are involved; in these cases one of the methodologies presented in the following sections needs to be utilized.

4.2.2   Artificial Neural Networks

Artificial Neural Networks are the main representatives of a more robust approach to classification; these classifiers, prior to being put to use, process available medical history in order to extract useful information concerning the underlying data patterns and structure, thus also acquiring the information required in order to optimize the classification process. Following a massively parallel architecture, quite similar to that of neurons in the human brain, ANNs construct powerful processing and decision-making engines through the combination of quite trivial processing units—also named neurons (Fig. 4.2).

In order to provide for some kind of control of the network operation and to facilitate training and fine-tuning, contrary to the case of networks of real neurons, connectivity in ANNs is usually constrained, in the sense that neurons are organized in layers and each neuron can

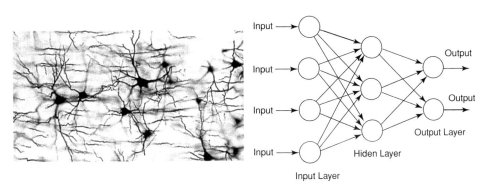

FIGURE 4.2: A network of real neurons compared to an artificial neural network

only accept inputs from the directly preceding layer and provide output to the directly following one.

Although ANNs are typically blamed by the scientific community for their nonsymbolic nature—i.e., the fact that their exact operation and the real meaning of their output is not easily revealed to the human inspector when their size grows—they are widely used in numerous classification problems, including CDSSs, due to their remarkable performance related characteristics. For example, the following works are based on applications of ANNs [11–15] while [16] reviews the benefits of ANN application in medicine in general.

The characteristic that makes ANNs so popular is the fact that given a set of labeled data (the medical history), an ANN will tune itself in an automated manner as to match these data in the best possible way. Unfortunately, this training process is not a trivial one. Numerous methodologies are presented in the literature for training ANNs, each one focusing on a different feature or situation. Thus, different training methodologies (as well as network structure) can be used when the number of training data is small, computing resources are limited, some or all of the historical medical data are unlabeled, there are missing data, the number of distinct patterns in the data is not known beforehand, and so on. For information on choosing, training, and using an ANN in a theoretical, research, or application environment, [17] is one of the best textbooks for both the novice and expert reader.

4.2.2.1 Clustering

One of the first things one finds out when trying to apply ANNs in a practical situation is the fact that their training is far from trivial. Even when having chosen the type of neurons, count of layers and training methodology to follow—neither of which is a trivial decision—the result of the training process is neither guaranteed nor set. Quite the contrary, depending on delicate issues such as the initial parameters of the network or even the order in which the training data are presented, the training procedure may differ in both time required and quality of results acquired. Clearly, it is desired to somehow decouple network performance from such issues. In order to avoid these dependences, one would need to have adequate prior knowledge as to initialize the ANN so that it is almost already trained correctly, in this case the adaptation taking place during training is minimal and does not depend that much on issues such as the order of the training data.

A common approach to this problem is to utilize a data-clustering approach in order to roughly extract the patterns that underlie in the data, and then map this information on the structure of the network. The results of this clustering process are so important for the training of the network that it is often mentioned that this may even be the most important determinant of the later operation of the CDSS as well [18].

Although a wide variety of clustering methodologies exists and has been used in the framework of ANN initialization, most methods fall under one of two categories: partitioning and agglomerative approaches. Partitioning approaches, much similarly to ANNs, start from some initial parameters and recursively optimize them. Specifically, given some initial estimation of the cluster's positions (centers, shapes, spreads, etc.) partitioning methods will classify all available data points to these clusters, and then use the results in order to update cluster parameters. Continuing recursively until equilibrium is found, changes are below a given threshold or enough repetitions have been made, the results of the process are refined. Variations, as for example having clusters with multiple centers, have also been proposed. Of course, in order for such a method to be applied, it is a requirement that available data can be positioned in an n-dimensional space.

Although partitioning approaches are the most robust and error resilient, the fact that they require an initial estimation as initialization makes them poor candidates for the initialization of ANNs, as they do not provide for complete automation of the process. Specifically, it is the fact that one needs to know the count of clusters, i.e., the count of distinct patterns in the data, beforehand that makes their application difficult.

On the other hand, agglomerative approaches start by considering each data sample as a distinct cluster and recursively merge the two most similar clusters until no similar clusters remain. The exact definition of the termination criterion as well as the definition of similarity between clusters are not trivial problems with unique answers; quite the contrary, they provide for a very wide variety of agglomerative clustering flavors.

The following works are based on the utilization of clustering approaches toward the initial exploration of the data patterns underlying in the available data [19–21], while [22] focuses specifically on the effect this has on the performance of the resulting ANN.

4.2.2.2 Self-Organizing Maps

Clearly, the fact that one needs to have a clear idea concerning the structure of the network (count of layers, count of neurons) before even training and testing can start is a very important limitation for ANNs. As one might expect, there have been efforts to produce ANN structures that overcome this. Resource-allocating neural networks were the first attempt in this direction. These networks have the ability to alter their structure during training by adding neurons when needed—and some times by even removing neurons no longer needed. Unfortunately, these networks remain sensitive to initial parameters, as incorrect initialization leads to the generation of too many neurons, thus resulting in over-fitting to the training data and poor classification performance when it comes to future data [22].

Kohonen's self-organizing maps (SOMs) constitute a more interesting and robust approach to training a network with no prior knowledge of the underlying data structures.

A number of details about the selection of the parameters, variants of the map, and many other aspects have been covered in the monograph [23]. SOMs, following an unsupervised learning approach, also achieve excellent data reduction and visualization, making it easier for humans to grasp the detected structures.

Due to their excellent properties and characteristics, SOMs have found numerous applications in CDSSs, such as [19, 24–27].

4.2.2.3 Support Vector Machines

Support vector machines (SVMs) are a relatively new type of training methodology for ANNs. Their characteristic is that they transform the input space, as to project training data from the original input space to a new one, where classes are linearly separable and thus linear learning algorithms can be applied. SVMs guarantee the maximal margin between different classes through global optimization of the decision boundary.

An important characteristic for SVM learning is that the margin between two classes depends solely on those data samples that are next to it. These training data samples—marked as gray in Fig. 4.3—are named support vectors and are the only ones to be considered when training the network (optimizing the margin between the classes).

Of course, once data have been projected to a space where they are linearly separable, training is both trivial and fast. Unfortunately, the choice of this mapping function, the kernel function, is not equally trivial. Prior knowledge of the characteristics of the data may contribute to a proper choice, but in practice selecting the optimal kernel is the main challenge in the application of SVMs [28].

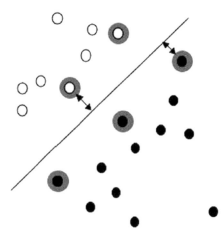

FIGURE 4.3: Operation of the support vector machine

Due to the fact that, unlike other ANNs, SVMs focus on maximizing the margin between classes, thus minimizing the probability of misclassification, they are extremely popular in CDSSs, where the cost of a misclassification may have a direct impact on human life. The following works are just a few examples of works in the medical field that are based on SVM learning [29–32].

4.2.3   Decision Trees

Humans using CDSSs, even those experts who are active in the research that develops them, are often reluctant to leave important medical decisions up to a subsymbolic, and thus generally incomprehensible, automated engine. A main drawback of ANNs in this framework is the fact that in order for a single output to be computed, numerous inputs are considered simultaneously and in different ways, which makes the process practically impossible for a human to follow. Decision trees offer an alternative computing methodology that reaches a decision through consecutive, simple question and answer sessions.

In the learning phase (when the decision tree is constructed) exactly one of the available features needs to be selected as the root feature; i.e. the most important feature in determining the diagnosis. Then data are split according to the value they have for this feature, and each group of data is used in order to create the corresponding child (subtree) of the root. If all of the data in a group belong to the same diagnosis, then that child becomes a leaf to the tree and is assigned that diagnosis. Otherwise, another feature is selected for that group, and data are again split leading to new groups and new children for this node. The choice of feature to use in order to split the data at any given step turns out to be a crucial one, as it can make the difference between a compact and a huge decision structure. Once a tree has been constructed in this manner, using the tree in order to classify a future case is quite similar to answering a small set of questions, each one regarding only one feature.

The decision tree in Fig. 4.4, for example, indicates that successful patients have more than one embryo transferred (NUM_TRANS > 1), have no additional hormonal stimulation (FLARE = 0), are of age less than or equal to 40 (AGE $\leq$ 40), and have no trauma during implantation of the embryos (TRAUMA = 0). Actually, in order to cope with errors in data, overlapping classes, and so on, real life decision trees also incorporate chance nodes, where instead of a clear decision, probabilities for different outcomes are presented.

Overall, decision trees are known for being robust to noisy data and capable of learning disjunctive expressions. Decision tree learning is one of the most widely used and practical methods for inductive inference. On the low side, the discrete structure of decision trees does not allow them to be applied in cases where the input variables are neither discrete nor easily partitioned in a small number of meaningful distinct ranges.

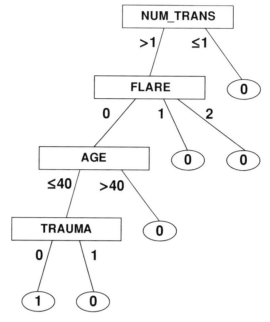

FIGURE 4.4: A sample decision tree

When it comes to CDSSs, decision trees were for long the most popular approach to decision making, and apparently, even though other data mining and machine learning approaches progress, still remain close to the top when it comes to popularity [33–37]. For a review of decision tree applications in medicine one may also consult [38]. It seems that decision trees will not fade out soon, as there is still work in progress toward further improving their performance, and their fields of application are constantly expanding.

4.2.4   Ensemble Machines

Classifiers such as the ones mentioned above, as well as many others, have found numerous applications in medicine and elsewhere. Each one of these classifiers finds a situation in which it performs its best, typically outperforming every other known approach. On the other hand, it is almost never possible to know beforehand which classifier will perform best on future data, which is the actual goal of designing and building a CDSS. In other words, although we have developed a powerful arsenal comprising a wide range of methodologies that one may use at any given time, we are quite unable to select the perfect tool for the job when it comes to developing a specific decision support system.

The ensemble methodology is the approach that helps overcome this obstacle. According to it, much similarly to having a group of doctors provide their independent opinions and

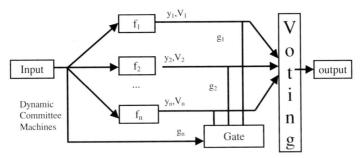

FIGURE 4.5: A dynamic committee machine architecture

then combining them to come up with a more reliable diagnosis, multiple simple classifiers are applied and their results are combined in order to acquire the output of the decision support system [39]. This combination can range from a simple averaging, to weighted averaging, or even to dynamic committee machines that consider the certainty in the output of each classifier, as estimated by the margin with which the classifier separates the classes for the specific inputs (Fig. 4.5).

As new information always reduces uncertainty, the consideration of more than one classifiers that have been trained and applied independently can only enhance the results of the classification process. Thus, ensemble approaches to classification allow for the generation of CDSSs with very high classification rates. For example, the following works have enhanced results due to the fact that they are based on ensemble approaches [40–42]. In general, it is now generally accepted that best results that one can acquire are those that come from the combination of multiple classifiers.

4.2.5   Fuzzy Logic and Fuzzy Rule Systems

In all classification schemes presented in the previous sections, automated analysis of medical history provided for the acquisition of the knowledge required for the operation of the CDSS, while the role of the human expert was limited to the labeling of historical medical data, i.e., to providing correct diagnoses for past cases. Alternatively to examining medical history in order to extract information, a CDSS may operate based on information already available to humans and provided directly to the system; in that case the role of the CDSS is to make up for human experts that cannot be available at all times when they may be needed.

There is typically no learning process in the development of a knowledge-based system. Instead, human experts provide directly the knowledge that the system needs to operate. Of course, this cannot be done at a subsymbolic level. The format of the knowledge needs to be one that facilitates the human experts; more so in knowledge-based systems developed for medical applications, since in this case the human experts are clinicians rather than knowledge engineers or computer scientists.

The systems developed based on the above principles, often also referred to as expert systems, typically use the form of rules such as the following in order to represent human knowledge:

IF *high_heart_rate* THEN *hyperthyroidism*

Although simple in definition, it turns out that these rules are not equally simple in evaluation. First of all, as can clearly be seen from the above example, measured medical features cannot be fed directly to the decision-making component; they first have to be transformed to lexical variables such as *has_euphoria*. One soon realizes that this cannot be done in a binary sense; for example there are different degrees of euphoria, as there are different degrees to which one may display manic symptoms. As a consequence, rule-based systems are closely related to the theory that best supports degrees and lexical variable, the theory of fuzzy sets and fuzzy logic.

In a fuzzy logic system, measured inputs are mapped to linguistic variables, as indicated in Fig. 4.6. A given measurement can be mapped to contradicting variables at the same time and to different degrees. As a result multiple, and often contradicting rules may fire, again each one to its degree. In order to acquire a classification, the result of this process is de-fuzzified; i.e. the activation levels of all rules are examined so that a crisp (not fuzzy) decision can be made. It is worth noting that, although the term "fuzzy rule" is quite common, the rules themselves are anything but fuzzy; it is the way they are being evaluated that associates them to degrees.

Due to the human perceivable nature of the knowledge representation format that rule-base systems use, it has always been a goal to design a rule-base system that is trained automatically from medical history. That would mean that the system, additionally to decision support, could offer the means to automatically explore data and propose possible associations to researchers. Neurofuzzy systems are one breed of classifiers that possess this property [17]. They have the structure and training characteristics of traditional ANNs, but their structure can be directly generated from, or be directly mapped to fuzzy rules. Unfortunately, the proper

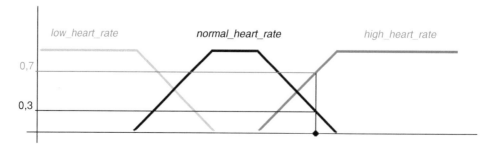

FIGURE 4.6: Three linguistic variables for the same measured feature

initialization and training of a neurofuzzy system turns out to be overly complex, as one needs to know beforehand how measure variables are mapped best to linguistic variables.

What seems to be more promising is the less direct combination of ANN operation with rule-base systems. As ANN-learning focuses on detecting underlying data patterns, if one is able to somehow extract the information stored in the structure and parameters of the network, then one can automatically generate the fuzzy rules that characterize the observed biological system; this process is known as rule extraction from ANNs. For some types of ANNs, as are, for example radial basis function (RBF) networks, the definition of the linguistic variable intervals, as well as the definition of the rules is not overly complex.

Still, the rules typically extracted in this process differ from what is desired, due to the fact that they are overly complex as they combine almost all available inputs. In order for both meaningful and human perceivable rules to be extracted, rule antecedents (inputs) that are not as important need to be trimmed; the ideal is to get rules that are as simple as possible and yet provide for a classification rate as high as that of the untrimmed rules. Various approaches have been proposed in this direction, with evolutionary ones being the ones that are currently gaining in momentum [43].

The fact that rule-base systems can i) accept information provided by human experts in a natural way and ii) provide the knowledge generated during their training in a human perceivable way has made them a very popular choice for CDSSs. As far as information extraction is concerned, there are numerous works focusing on rule extraction from ANNs that have been trained on medical data [44]. As far as the application of fuzzy logic and rule-base systems in a medical context, the following are only few examples [45, 46]. For a review of the field, readers may refer to [47, 48]. When it comes to actual implemented systems, they are far too many to mention. For an indicative list of more than 60 real life systems, one may refer to Federhofer's page [49].

4.3   EVALUATION

Although all of the methodologies and approaches presented in this chapter have found applications in the medical field with important results, none of them is a panacea that is suitable for utilization in all cases. Quite the contrary, any of these methodologies will perform at a level much below what one might consider as acceptable when not properly initialized or fine-tuned to the problem at hand. As a result, before a CDSS is put into actual use, it needs to be carefully evaluated, so that its error margins and weaknesses in general are identified and clinicians are informed about them; obviously, a clinician should not weigh the diagnosis of an automated system with 99% accuracy similarly to that of a system with 85% accuracy.

In classification research in general, a system is both trained and evaluated using available medical history. When the data suffice in number, a part of it (typically the smaller part) is used for training, while remaining data are used for testing. When less data are available, techniques such as leave one out may be used, where all but one available data points are used to train the classifier and the remaining point is used for testing; all points recursively play the role of the testing data and overall results are evaluated statistically [5].

While both approaches are meaningful from the evaluator's point of view, they are not necessarily 100% reasonable for a CDSS. As far as the leave one out approach is concerned, it provides an estimation of the potential of the overall classification methodology applied; as multiple distinct classifiers are trained, one cannot tell at the end which classifier parameters the results refer to, in order to construct the CDSS that can achieve similar classification rates. As far as training data/testing data separation is concerned, using the smaller part of the data for training means that useful information is not considered in the construction of the CDSS in order to pursue maximum performance. One can see that the way to evaluate a CDSS is a challenge on its own, often exceeding in difficulty and importance the development of the system itself. Typically, a training data/testing data approach is followed, but with most of the available medical history used for training and a smaller part as test data.

A more subtle but equally important issue to consider is that of the suitability of the medical history for the evaluation of the system. Clinicians now accept that living in different regions of the world, as well as living in different climates and so on, affects one's body in a way that may make different medical feature measurements qualify as "normal" in each case. As a result, medical history collected from patients from a specific part of the world may not be suitable in order to evaluate a system that will be used in another. Additionally, medical history of previous years may not always be suitable for the evaluation of a system that will be put in use today. It is exactly this last remark that makes the evaluation of CDSSs an ongoing process; a CDSS that has been optimally configured may need to be reconfigured in order to adapt to an era where a difference in feeding habits or even simply in lifestyle has an effect on typically measured medical values, when clinicians empirically observe that error margins have augmented.

From a technical point of view, an interesting point to consider is the way to measure the performance of a CDSS. For classification systems in general, evaluation is typically based on a true/false and positive/negative scheme. When adopted in the medical case, true positive (TP) is a correct illness estimation, true negative (TN) a correct healthy estimation, false positive (FP) an illness estimation for a healthy case, and a false negative (FN) a healthy estimation for an ill case. Based on these, accuracy is defined as follows:

$$Accuracy = \frac{TP + TN}{TP + TN + FP + FN} \qquad (4.1)$$

The simplistic approach of simply counting correct and incorrect classifications in order to estimate accuracy, although generally accepted in other expert systems and classifiers, is not sufficient for the case of medical systems, where one type of mistake may be much more important—as far as the possible consequences are concerned—compared to another. For example a false positive estimation has a result of a patient taking extra tests in order to verify one's health status, whereas a false negative diagnosis may deprive one from an early diagnosis and treatment. Finally, classes in a medical setting are rarely balanced; it is only typical that only a small percentage of people examined will actually be ill. As a result, a system that always provides a "healthy" diagnosis reaches high classification rates.

In order to compensate for this, a more flexible consideration of errors needs to be used, in order for class probabilities to be considered as well. A simple approach that is commonly followed in this direction is the utilization of specificity and sensitivity measures, defined as follows:

$$Specificity = \frac{TN}{TN + FP} \tag{4.2}$$

$$Sensitivity = \frac{TP}{TP + FN} \tag{4.3}$$

where specificity and sensitivity are actually measures of accuracy, when considering only healthy or only ill cases, respectively, thus decoupling the measures from class probabilities.

4.4   CONCLUSIONS

CDSSs originate from data mining and generic decision support systems, both of which are fields that have matured over the years and are now at a position to provide reliable and robust practical solutions. As a result, high-quality CDSSs have been developed. Still, one cannot forget that an automated decision support system cannot become perfect, when the data it relies its operation on is not.

It is perhaps not surprising that numerous clinicians hesitate to utilize decision support systems in their line of work [50]. This is of course disappointing, given the fact that CDSSs when properly used have been found to make a clear difference in the quality of treatment patients receive [51]. Lately, the opposite, and quite possible more dangerous, extent is also observed at times, with clinicians often trusting CDSSs more than their own judgment [52].

Clearly, the goal is to reach a stage where CDSSs are integrated in the process of everyday clinical work, but without being assigned roles they are not made for, such as the role of the actual clinician. It seems that a number of parameters will have an effect in this process, ranging from purely financial issues to degree of automation, from availability at the time and location support is needed to the ease of the user interface, and from the quality of the data acquisition components to the success of the system development and integration procedures [53]. The areas in which a CDSS could make a difference are, of course, countless. Table 4.1 provides only a summary of the most important ones.

TABLE 4.1: Potential Users and Uses for a CDSS

USER	APPLICATION
Pharmacists	Drug levels, drug/drug interactions, culture and sensitivity results, adverse drug events
Physicians	Critical lab results, culture and sensitivities
Nurses	Critical lab results, drug/drug interactions
Dietary	Patient transfers, lab support for tube feedings
Epidemiology/infection control	VRE and MRSA results, reportable organisms
Housekeeping	ADT functions
Billing	Excessively expensive tests and treatments
Administration	Patient chart administration
Chaplain	Counseling opportunity
Case manager	Premature readmission
Patient	Drug/drug interactions, drug dosing, missing tests

Regardless of their penetration degree in traditional clinical practice, CDSSs have not yet reached the extent of application one might hope for in the telemedicine case. In this framework, e-Health continues to refer mainly to distributed information systems for the organization and exchange of medical records and information in general; automated processing and decision support components are still considered by many as a luxury for e-Health, mainly due to their augmented networking and processing needs.

As networking bandwidth is now becoming cheaper and more easily available, and portable and handheld devices are becoming more powerful, it is only natural to expect a growth in the number and type of services e-Health will incorporate, with decision support being a primary candidate. Already a number of Internet-based or distributed medical decision support systems have been proposed and developed [54–56], while others are certainly on their way.

REFERENCES

[1] L. E. Perreault, and J. B. Metzger, "A pragmatic framework for understanding clinical decision support," *J. Healthc. Inf. Manage. Syst. Soc.*, vol. 13, no. 2, pp. 5–21, 1999.

[2] J. C. Wyatt and J. L. Liu, "Basic concepts in medical informatics," *J. Epidemiol. Community Health*, vol. 56, pp. 808–812, 2002.doi:10.1136/jech.56.11.808

[3] B. Kaplan, "Evaluating informatics applications—clinical decision support systems lit-erature review," *Int. J. Med. Inf.*, vol. 64, no. 1, pp. 15–37, 2001.doi:10.1016/S1386-5056(01)00183-6

[4] K. Zheng, R. Padman, M. P. Johnson, and H. S. Diamond, "Understanding technology adoption in clinical care: Clinician adoption behavior of a point-of-care reminder system," *Int. J. Med. Inf.*, vol. 74, no. 7–8, pp. 535–543, 2005.doi:10.1016/j.ijmedinf.2005.03.007

[5] T. S. Lim, W. Y. Loh, and Y. S. Shih, "A Comparison of prediction accuracy, complexity, and training time of thirty-three old and new classification algorithms," *Mach. Learn.*, vol. 40, pp. 203–229, 2000.doi:10.1023/A:1007608224229

[6] L. Bull, *Applications of Learning Classifier Systems*. New York: Springer-Verlag, 2004.

[7] M. Ohlsson, "WeAidU—A decision support system for myocardial perfusion images using artificial neural networks," *Artif. Intell. Med.*, vol. 30, no. 1, pp. 49–60, 2004. doi:10.1016/S0933-3657(03)00050-2

[8] M. Hilario, A. Kalousis, M. Muller, and C. Pellegrini, "Machine learning approaches to lung cancer prediction from mass spectra," *Proteomics*, vol. 3, pp. 1716–1719, 2003. doi:10.1002/pmic.200300523

[9] J. Prados, A. Kalousis, J. C. Sanchez, L. Allard, O. Carrette, and M. Hilario, "Mining mass spectra for diagnosis and biomarker discovery of cerebral accidents," *Proteomics*, vol. 4, pp. 2320–2332, 2004.doi:10.1002/pmic.200400857

[10] M. Wagner, D. Naik, A. Pothen, S. Kasukurti, R. Devineni, B. L. Adam, O. J. Semmes, and G. L. Wright Jr., "Computational protein biomarker prediction: A case study for prostate cancer," *BMC Bioinform.*, vol. 5, p. 26, 2004.

[11] A. E. Smith, C. D. Nugent, and S. I. McClean, "Evaluation of inherent performance of intelligent medical decision support systems: Utilising neural networks as an example," *Artif. Intell. Med.*, vol. 27, no. 1, pp. 1–27, 2003.doi:10.1016/S0933-3657(02)00088-X

[12] M. E. Futschik, M. Sullivan, A. Reeve, and N. Kasabov, "Prediction of clinical be-haviour and treatment for cancers," *OMJ Appl. Bioinform.*, vol. 2, no. 3, pp. 53–58, 2003.

[13] G. Ball, S. Mian, F. Holding, R. O. Allibone, J. Lowe, S. Ali, G. Li, S. McCardle, I. O. Ellis, C. Creaser, and R. C. Rees, "An integrated approach utilizing artificial neural networks and SELDI mass spectrometry for the classification of human tumours and rapid identification of potential biomarkers," *Bioinformatics*, vol. 18, no. 3, pp. 395–404, 2002.doi:10.1093/bioinformatics/18.3.395

[14] L. J. Lancashire, S. Mian, I. O. Ellis, R. C. Rees, and G. R. Ball, "Current developments in the analysis of proteomic data: Artificial neural network data mining techniques for the identification of proteomic biomarkers related to breast cancer," *Curr. Proteom.*, vol. 2, no. 1, pp. 15–29, 2005.doi:10.2174/1570164053507808

[15]   I. Maglogiannis and E. Zafiropoulos, "Utilizing support vector machines for the characterization of digital medical images," *BMC Med. Inf. Decis. Making*, vol. 4, no. 4, 2004, http://www.biomedcentral.com/1472-6947/4/4/

[16]   P. J. Lisboa, "A review of evidence of health benefit from artificial neural networks in medical intervention," *Neural Netw.*, vol. 15, no. 1, pp. 11–39, 2002.doi:10.1016/S0893-6080(01)00111-3

[17]   S. Haykin, *Neural Networks: A Comprehensive Foundation*, 2nd ed. Prentice Hall, 1999.

[18]   J. H. Chuang, G. Hripcsak, and R. A. Jenders, "Considering clustering: A methodological review of clinical decision support system studies," in *Proc. AMIA Symp.*, 2000, pp. 146–150.

[19]   H. Wang, F. Azuaje, and N. Black, "An integrative and interactive framework for improving biomedical pattern discovery and visualization," *IEEE Trans. Inf. Technol. Biomed.*, vol. 8, no. 1, pp. 16–27, 2004.doi:10.1109/TITB.2004.824727

[20]   H. M. Kuerer, K. R. Coombes, J. N. Chen, L. Xiao, C. Clarke, H. Fritsche, S. Krishnamurthy, S. Marcy, M. C. Hung, and K. K. Hunt, "Association between ductal fluid proteomic expression profiles and the presence of lymph node metastases in women with breast cancer," *Surgery*, vol. 136, no. 5, pp. 1061–1069, 2004. doi:10.1016/j.surg.2004.04.011

[21]   P. V. Purohit and D. M. Rocke, "Discriminant models for high-throughput proteomics mass spectrometer data," *Proteomics*, vol. 3, pp. 1699–1703, 2003. doi:10.1002/pmic.200300518

[22]   M. Wallace, N. Tsapatsoulis, and S. Kollias, "Intelligent initialization of resource allocating RBF networks," *Neural Netw.*, vol. 18, no. 2, pp. 117–122, 2005. doi:10.1016/j.neunet.2004.11.005

[23]   T. Kohonen, *Self-Organizing Maps*, 2nd ed. Berlin: Springer-Verlag, 1997.

[24]   T. P. Conrads, V. A. Fusaro, S. Ross, D. Johann, V. Rajapakse, B. A. Hitt, S. M. Steinberg, E. C. Kohn, D. A. Fishman, G. Whitely, J. C. Barrett, L. A. Liotta, E. F. Petricoin, and T. D. Veenstra, "High-resolution serum proteomic features for ovarian cancer detection," *Endocr. Relat. Cancer*, vol. 11, no. 2, pp. 163–178, 2004.doi:10.1677/erc.0.0110163

[25]   D. J. Johann Jr., M. D. McGuigan, S. Tomov, V. A. Fusaro, S. Ross, T. P. Conrads, T. D. Veenstra, D. A. Fishman, G. R. Whiteley, E. F. Petricoin, and L. A. Liotta, "Novel approaches to visualization and data mining reveals diagnostic information in the low amplitude region of serum mass spectra from ovarian cancer patients," *Dis. Markers*, vol. 19, pp. 197–207, 2004.

[26]   D. Ornstein, W. Rayford, V. Fusaro, T. Conrads, S. Ross, B. Hitt, W. Wiggins, T. Veenstra, L. Liotta, and E. Petricoin, "Serum proteomic profiling can discriminate prostate cancer from benign prostates in men with total prostate specific antigen

levels between 2.5 and 15.0 NG/ML," *J. Urol.*, vol. 172, no. 4, pp. 1302–1305, 2004.doi:10.1097/01.ju.0000139572.88463.39

[27]   J. H. Stone, V. N. Rajapakse, G. S. Hoffman, U. Specks, P. A. Merkel, R. F. Spiera, J. C. Davis, E. W. St.Clair, J. McCune, S. Ross, B. A. Hitt, T. D. Veenstra, T. P. Conrads, L. A. Liotta, and E. F. Petricoin, "A serum proteomic approach to gauging the state of remission in Wegener's granulomatosis," *Arthritis Rheum.*, vol. 52, pp. 902–910, 2005. doi:10.1002/art.20938

[28]   H. Shi and M. K. Markey, "A machine learning perspective on the development of clinical decision support systems utilizing mass spectra of blood samples," *J. Biomed. Inf.*, to be published.

[29]   A Statnikov, C. F. Aliferis, and I. Tsamardinos, "Methods for multi-category cancer diagnosis from gene expression data: A comprehensive evaluation to inform decision support system development," *Medinfo*, vol. 11, pp. 813–817, 2004.

[30]   L. Li, H. Tang, Z. Wu, J. Gong, M. Gruidl, J. Zou, M. Tockman, and R. A. Clark, "Data mining techniques for cancer detection using serum proteomic profiling," *Artif. Intell. Med.*, vol. 32, pp. 71–83, 2004.doi:10.1016/j.artmed.2004.03.006

[31]   B. Wu, T. Abbott, D. Fishman, W. McMurray, G. Mor, K. Stone, D. Ward, K. Williams, and H. Zhao, "Comparison of statistical methods for classification of ovarian cancer using mass spectrometry data," *Bioinformatics*, vol. 19, no. 13, pp. 1636–1643, 2003. doi:10.1093/bioinformatics/btg210

[32]   I. Maglogiannis, S. Pavlopoulos, and D. Koutsouris, "An integrated computer supported acquisition, handling and characterization system for pigmented skin lesions in derma-tological images," *IEEE Trans. Inf. Technol. Biomed.*, vol. 9, no. 1, pp. 86–98, 2005. doi:10.1109/TITB.2004.837859

[33]   J. R. Trimarchi, J. Goodside, L. Passmore, T. Silberstein, L. Hamel, and L. Gonzalez, "Assessing decision tree models for clinical in-vitro fertilization data," Dept. of Computer Science and Statistics, Univ. Rhode Island, Tech. Rep. TR03-296, 2003.

[34]   R. D. Niederkohr and L. A. Levin, "Management of the patient with suspected temporal arteritis: A decision-analytic approach," *Ophthalmology*, vol. 112, no. 5, pp. 744–756, 2005.doi:10.1016/j.ophtha.2005.01.031

[35]   N. Ghinea and J. M. Van Gelder, "A probabilistic and interactive decision-analysis system for unruptured intracranial aneurysms," *Neurosurg. Focus*, vol. 17, no. 5, p. E9, 2004.

[36]   M. K. Markey, G. D. Tourassi, and C. E. J. Floyd, "Decision tree classification of proteins identified by mass spectrometry of blood serum samples from people with and without lung cancer," *Proteomics*, vol. 3, no. 9, pp. 1678–1679, 2003.doi:10.1002/pmic.200300521

[37]   H. Zhu, C. Y. Yu, and H. Zhang, "Tree-based disease classification using protein data," *Proteomics*, vol. 3, no. 9, pp. 1673–1677, 2003.doi:10.1002/pmic.200300520

[38]   V. Podgorelec, P. Kokol, B. Stiglic, and I. Rozman, "Decision trees: An overview and their use in medicine," *J. Med. Syst.*, vol. 26, no. 5, pp. 445–463, 2002. doi:10.1023/A:1016409317640

[39]   G. Fumera and F. Roli, "A theoretical and experimental analysis of linear combiners for multiple classifier systems," *IEEE Trans. Pattern Anal. Mach. Intell.*, vol. 27, no. 6, pp. 942–956, 2005.doi:10.1109/TPAMI.2005.109

[40]   P. Mangiameli, D. West, and R. Rampal, "Model selection for medical diagnosis decision support systems," *Decis. Support Syst.*, vol. 36, no. 3, pp. 247–259, 2004. doi:10.1016/S0167-9236(02)00143-4

[41]   A. C. Tan and D. Gilbert, "Ensemble machine learning on gene expression data for cancer classification," *Appl. Bioinform.*, vol. 2, no. 3, pp. 75–83, 2003.

[42]   P. L. Martelli, P. Fariselli, and R. Casadio, "An ENSEMBLE machine learning approach for the prediction of all-alpha membrane proteins," *Bioinformatics*, vol. 19, pp. 205–211, 2003.doi:10.1093/bioinformatics/btg1027

[43]   M. Wallace and N. Tsapatsoulis, "Combining GAs and RBF neural networks for fuzzy rule extraction from numerical data," in *Int. Conf. Artificial Neural Networks*, Warsaw, Poland, Sep. 2005.

[44]   J. R. Rabunal, J. Dorado, A. Pazos, J. Pereira, and D. Rivero, "A new approach to the extraction of ANN rules and to their generalization capacity through GP," *Neural Comput.*, vol. 16, no. 7, pp. 1483–1523, 2004.doi:10.1162/089976604323057461

[45]   J. H. T. Bates and M. P. Young, "Applying fuzzy logic to medical decision making in the intensive care unit," *Am. J. Respir. Crit. Care Med.*, vol. 167, no. 7, pp. 948–952, 2003. doi:10.1164/rccm.200207-777CP

[46]   P. R. Innocent, R. I. John, and J. M. Garibaldi, "Fuzzy methods for medical diagnosis," *Appl. Artif. Intell.*, vol. 19, no. 1, pp. 69–98, 2005.doi:10.1080/08839510590887414

[47]   S. Barro and R. Marin, *Fuzzy Logic in Medicine*. Physica-Verlag (Heidelberg), 2002.

[48]   M. Mahfouf, M. F. Abbod, and D. A. Linkens, "A survey of fuzzy logic monitoring and control utilisation in medicine," *Artif. Intell. Med.*, vol. 21, no. 1–3, pp. 27–42, 2001. doi:10.1016/S0933-3657(00)00071-3

[49]   http://www.computer.privateweb.at/judith/special_field3.htm

[50]   N. Rousseau, E. McColl, J. Newton, J. Grimshaw, and M. Eccles, "Practice based, longitudinal, qualitative interview study of computerized evidence-based guidelines in primary care," *Br. Med. J.*, vol. 326, pp. 1–8, 2003.

[51]   A. X. Garg, N. K. Adhikari, H. McDonald, M. P. Rosas-Arellano, P. J. Devereaux, J. Beyene, J. Sam, and R. B. Haynes, "Effects of computerized clinical decision support systems on practitioner performance and patient outcomes: A systematic review," *J. Am. Med. Assoc.*, vol. 293, no. 10, pp. 1223–1238, 2005.doi:10.1001/jama.293.10.1223

[52] S. Dreiseitl and M. Binder, "Do physicians value decision support? A look at the effect of decision support systems on physician opinion," *Artif. Intell. Med.*, vol. 33, no. 1, pp. 25–30, 2005.doi:10.1016/j.artmed.2004.07.007

[53] K. Kawamoto, C. A. Houlihan, E. A. Balas, and D. F. Lobach, "Improving clinical practice using clinical decision support systems: A systematic review of trials to identify features critical to success," *Br. Med. J.*, vol. 330, no. 7494, pp. 740–741, 2005.

[54] D. Jegelevicius, V. Marozas, A. Lukosevicius, and M. Patasius, "Web-based health services and clinical decision support," *Stud. Health Technol. Inform.*, vol. 105, pp. 27–37, 2004.

[55] P. Fortier, S. Jagannathan, H. Michel, N. Dluhy, and E. Oneil, "Development of a handheld real-time decision support aid for critical care nursing," *HICSS'03*, vol. 6, no. 6, p. 160, 2003.

[56] G. Eysenbach, J. Powell, M. Englesakis, C. Rizo, and A. Stern, "Health related virtual communities and electronic support groups: Systematic review of the effects of online peer to peer interactions," *Br. Med. J.*, vol. 328, no. 7449, pp. 1–6, 2004.

CHAPTER 5

Medical Data Coding and Standards

5.1   INTRODUCTION

Standards are generally required when excessive diversity creates inefficiencies or impedes effectiveness. Such a case exists in the electronic healthcare environment. In healthcare, standards are needed for encoding data about the patient that are collected by one system and used by another. One obvious need is the development of standardized identifiers for individuals, healthcare providers, health plans, and employers so that such individuals can be recognized across systems. Because of the diversity of patient data creation and storage points and the variety of medical IT-based equipment, healthcare is one area that can ultimately greatly benefit from implementation of standards [1].

Patients receive care across primary, secondary, and tertiary care settings with little bidirectional communication and coordination among the services. Patients are cared for by one or more primary physicians as well as by specialists. There is little coordination and sharing of data between inpatient and outpatient care. Within the inpatient setting, the clinical environment is divided into clinical specialties that frequently treat the patient without regard to what other specialties have done [1].

The lack of "standards" for electronic coding of medical data has been a major obstacle to establishing an integrated electronic medical record. However, during the last years significant efforts by major standard bodies (CEN, IEEE, HL7, ANSI, ISO) have produced several standards in medical informatics; some of them (i.e., HL7 and DICOM) are well established in the e-health sector. This chapter provides short descriptions of the dominant standards, indicating how they may be used for the development of distributed networked e-health applications.

5.2   THE MAJOR MEDICAL INFORMATICS CODING STANDARDS

5.2.1   Health Level 7 (HL7)

HL7 stands for Health Level 7 and is the highest level of the ISO communication model. It is a standard for data interchange. The model's purpose is to archive OSI (open systems interconnections), a way to get different systems to work together. The OSI model is not specific to medical informatics; however, HL7 is specific to the healthcare domain.

The current version of HL7 is version 3 with basic features being the object-oriented approach, the definition of a *Reference Information Model* (*RIM*) in UML defining all messages, objects and their relationships, the inclusion of specific conformance to the standard statements, and the use of *extensive markup language* (*XML*) as a structured communication language.

The use of the XML and the RIM has enabled the exchange of medical data among different healthcare organizations and devices and consequently the development of distributed networked e-health applications. For instance external systems with XML-aware browsers can communicate with a hospital information system (HIS) through the HL7 standard.

The structure of an HL7 message is as follows:

- Each message is constituted by specific **segments** in a concrete line.

- Each message has a **message type** that determines his scope.

- Each type of message corresponds in one or more actual incidents (**trigger event**). Actual incidents can be, for example, a patient admission or a medical examination order.

- Each segment has specific **data fields**.

- Segments in an HL7 message are either mandatory or optional while it is possible they are repeated in a message. Each segment has a concrete name. For example a message related to patient admission discharge and transfer functions (Type: ADT) can include the following segments: Message Header (MSH), Event Type (EVN), Patient ID (PID), and Patient Visit (PV1)

Such a message produced to notify the admission to surgery of a patient is as follows:

- MSH|**ADT1**|MCM|**LABADT**|MCM|**199808181126**|SECURITY|ADT^A01| MSG00001|**P**|**3.0**|

- EVN|A01|**199808181123**||

- PID|||PATID1234^5^M11|**JONES^WILLIAM^A^III**||19610615|M||C|1200 N ELM STREET^^**GREENSBORO**^NC^27401-1020|GL|(919)379-1212| (919)271434||S||PATID**12345001**^2^M10|123456789|987654^NC|

- NK1|**JONES^BARBARA^K|WIFE**||||||NK^NEXT OF KIN

- PV1|1|I|**2000**^**2012**^**01**|||||**004777^LEBAUER^SIDNEY**^J.|||**SUR**|||| ADM| A0|

The message explanation is as follows: Patient William A. Jones, III was admitted for surgery (SUR) on 18th July 1998, 11.23 p.m. from the doctor Sidney J. Lebauer (# 004777).

He entered the surgical room 2012, bed 01 in the nursing unit 2000. This message was sent by the system ADT 1 to the system LABADT, 3 minutes afterward admission.

5.2.2   International Classification of Diseases– (ICD)

ICD is a specific standard for medical terminologies. It was first published in 1893 and has been revised at roughly every 10 year. The revisions are underlaid the World Health Organization (WHO). The tenth edition was published in 1992. The coding consists of a core classification of three digits. A fourth digit (the decimal) is used for further detail. The terms are arranged in a hierarchy based on the digits.

The ICD has become the international standard diagnostic classification for all general epidemiological and many health management purposes. These include the analysis of the general health situation of population groups, and monitoring of the incidence and prevalence of diseases, and other health problems in relation to other variables such as the characteristics and circumstances of the individuals affected.

It is used to classify diseases and other health problems recorded on many types of health and vital records including death certificates and hospital records. In addition to enabling the storage and retrieval of diagnostic information for clinical and epidemiological purposes, these records also provide the basis for the compilation of national mortality and morbidity statistics by WHO member states.

ICD-10 uses alphanumeric codes, with an alphabetic character in the first position. It includes the letters "I" and "O," which are usually avoided in alphanumeric coding because of the likelihood of confusion with the numerals "1" and "0". Somewhat surprising is the fact that chapters in *ICD-10* do not necessarily start with new alphabetic characters. For example see Table 5.1.

ICD is incorporated in most of electronic health record (EHR) implementations; therefore, it is quite powerful tool for the development of integrated distributed health records, concentrating the total medical history of a patient. Further information regarding the ICD standard is out of the scope of this textbook and may be found in [8].

5.2.3   Digital Imaging and Communications in Medicine (DICOM)

DICOM (DICOM homepage URL: http://medical.nema.org/) is a standard for exchanging medical images. Medical images comply with the DICOM format [4]. A single DICOM file contains both a header (which stores information about the patient's name, the type of scan, image dimensions, etc.) and all of the image data. The header contains information regarding its size, the size of the whole image, and usually information that has to do with the image type and characteristics (resolution, possible compression, number of frames, etc). Table 5.2 shows a DICOM compliant image file representation, where after the header section the image data follow.

The standard also specifies the following:

TABLE 5.1: An Example of an ICD-10 Classification Table

Chapter I	Certain infectious and parasitic diseases	A00–B99
Chapter II	Neoplasms	C00–D48
Chapter III	Diseases of the blood and blood-forming organs and certain disorders involving the immune mechanism	D50–D89
Chapter IV	Endocrine, nutritional, and metabolic diseases	E00–E90
. . .		
Chapter IX	Diseases of the circulatory system	I00–I99
. . .		
Chapter XV	Pregnancy, childbirth, and the puerperium	O00–O99
. . .		

1. A set of protocols for devices communicating over a network.

2. The syntax and semantics of commands and associated information that can be exchanged using these protocols.

TABLE 5.2: DICOM Compliant Image File Format Representation

← Header →			← Image →
Transfer Syntax UID	**Definition**	**Value**	
0002,0000	File meta elements group length	150	
0002,0001	File meta info version	256	
0008,0008	Image type	SINGLE PLANEA	
0008,0020	Study date	19941013	
0008,0030	Study time	141917	
0010,0030	Patient date of birth	19751025	
0010,0040	Patient sex	M	
0028,0004	Photometric interpretation	MONOCHROME2	
0010,0020	Patient ID	556342B	
0008,0090	Referring physician's name	Dr. A	
0008,0080	Institution name	Medical Center	
0010,0010	Patient's name	Patient A	

3. A set of media storage services and devices claiming conformance to the standard, as well as a file format, and a medical directory structure to facilitate access to the images and related information stored on media that share information.

The DICOM standard is used in almost all e-health applications handling medical images (i.e, picture-archiving and communicating systems, PACS, and telemedicine and teleradiology).

5.2.4   Other Standards

ASTM (ASTM homepage. URL: http://www.astm.org/) is a standard for message exchange about clinical observations, medical logic, and electrophysiologic signals.

ADA is a standard for data exchange and processing in the dental health care sector.
ANA is a standard for data exchange and processing for nursing services.
LOINC (The logical observation identifiers names and codes) provides a standard for a set of universal names and codes, clinical observations, and diagnostic study observations.
EDIFACT is a set of standards for interchange of data between independent computer-based systems.

5.3   BIOSIGNAL CODING STANDARDS

Biosignals in networked e-health applications are provided in numerical format (i.e., heart rate, respiratory rate, temperature of body, arterial blood pressure, etc.) and as waveforms (Electro-cardiogram, ECG, and Electroencephalogram, EEG). Numerical data are very easy to handle and code in applications. Regarding the waveforms the acquisition and transmission of ECG is the most usual technique for the determination and the verification of a patient's condition in a remote position. Almost all modern ECG (mobile/wearable or not) are equipped with digital outputs and use digital techniques for communication. The computer-based coding and communication of ECG is regulated by the Standard Communications Protocol, SCP-ECG. SCP-ECG is a file format for ECG traces, annotations, and metadata. It is defined in the joint ANSI/AAMI standard EC71:2001 and in the CEN standard EN 1064:2005.

Practical experience during implementation and in the field confirmed its usability for telemetric applications as well as for data volume effective storage and retrieval (demonstrated in the OEDIPE project [9]).

5.3.1   SCP-ECG Standard Description

SCP-ECG is similar to the DICOM formatting concept. The ECG data are divided into different sections. Contents and format of each section are defined in the SCP document [10]. A global overview of the SCP-ECG data structure is presented in Table 5.3.

TABLE 5.3: Overview of the SCP-ECG Data Structure

CRC CHECKSUM
SIZE OF THE ENTIRE SCP-ECG RECORD
POINTERS TO DATA AREAS IN THE RECORD ("Table of content")
HEADER, e.g., PatID, Device ID, Recording ID (Time stamp, etc.)
ECG Data in optional formats without/with (selectable) compression methods
Various types of processing and over reading results

More specifically the information provided above is included in the SCP standard shown in Table 5.4 [10].

Each section is divided into two parts: The section ID Header and the section Data Part (see Fig. 5.1).

Although the section ID Header always has a length of 16 bytes, the section data part is variable. The SCP standard allows for a rather large number of options to store and format the ECG data. ECG data may be acquired at different sampling rates, with different quantization levels; they may be not compressed or be compressed by selectable methods and an SCP-ECG record may or may not contain analysis and over reading results. Also, the number of leads, the length of the recording interval and even the simultaneity of leads is left open to the manufacturers.

TABLE 5.4: SCP-ECG Data Fields	
Mandatory	2 BYTES—CHECKSUM—CRC—CCITT OVER THE ENTIRE RECORD (EXCLUDING THIS WORD)
Mandatory	4 BYTES—(UNSIGNED) SIZE OF THE ENTIRE ECG RECORD (IN BYTES)
Mandatory	(Section 0)
	POINTERS TO DATA AREAS IN THE RECORD
Mandatory	(Section 1)
	HEADER INFORMATION—PATIENT DATA/ECG ACQUISITION DATA
Optional	(Section 2)
	HUFFMAN TABLES USED IN ENCODING OF ECG DATA (IF USED)

TABLE 5.4: (*Continued*)

Optional	(Section 3) ECG LEAD DEFINITION
Optional	(Section 4) QRS LOCATIONS (IF REFERENCE BEATS ARE ENCODED)
Optional	(Section 5) ENCODED REFERENCE BEAT DATA IF REFERENCE BEATS ARE STORED

Optional	(Section 6)
	"RESIDUAL SIGNAL" AFTER REFERENCE BEAT
	   SUBTRACTION IF REFERENCE BEATS ARE STORED,
	   OTHERWISE ENCODED RHYTHM DATA

Optional	(Section 7) GLOBAL MEASUREMENTS
Optional	(Section 8) TEXTUAL DIAGNOSIS FROM THE "INTERPRETIVE" DEVICE
Optional	(Section 9) MANUFACTURER SPECIFIC DIAGNOSTIC AND OVERREADING DATA FROM THE "INTERPRETIVE" DEVICE
Optional	(Section 10) LEAD MEASUREMENT RESULTS
Optional	(Section 11) UNIVERSAL STATEMENT CODES RESULTING FROM THE INTERPRETATION

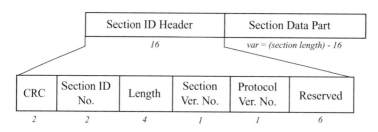

FIGURE 5.1: SCP-ECG section structure.

REFERENCES

[1] W. E. Hammond, J. J. Cimino, "Standards in medical informatics (Chapter 6)," in *Medical Informatics: Computer Applications in Health Care and Biomedicine*, E. H. Shortliffe and L.E. Perreault, Eds. New York: Springer-Verlag, 2000.

[2] http://www.journeyofhearts.org/jofh/jofh_old/minf_528/ intro.htm

[3] K. W. Goodman, *Ethics, Computing, and Medicine. Information and the Transformation of Health Care*. Cambridge: Cambridge University Press, 1998.

[4] http://www.deepthought.com.au/health/HIS_manifesto/Output/his_manifesto.html, Thomas Beale.

[5] http://www.diffuse.org/medical.html

[6] www.dicom.org

[7] www.hl7.org

[8] http://www.who.int/classifications/icd/en/

[9] www.ehto.org/aim/volume2/oedipe.html

[10] http://www.openecg.net/

CHAPTER 6

Web Technologies for Networked E-Health Applications

6.1   INTRODUCTION

When networking was first introduced in the field of medical applications, its role was limited to facilitating the connectivity and exchange of data between a central data center and various terminals in the scope of a clinical facility. In those days the resources provided by any trivial and proprietary local area network were more than sufficient for the development of the desired applications. Nowadays that networked e-Health applications of much larger scale and more importantly much broader reach are produced, the scope of medical applications has by far surpassed that of local area networks [1]. As one might expect, rather than reinventing the wheel, for reasons related not only to cost but mainly to interoperability, current and emerging networked e-Health applications rely on the existing networking infrastructure and standards, i.e., the Internet and its protocols, in order to reach its users.

There have been at times, and still are, barriers of various natures for the utilization of the Web and its technologies in a medical framework. For example, parameters such as computer literacy and Internet access greatly affect the degree to which the existence of Web-based applications provides for augmented accessibility to health services [2]. Ethical issues also have a major role in the development of Web-based medical applications [3]. Still the utilization of Web technologies in telemedicine and e-Health has already reached remarkable extents and continues to grow. In this chapter we review the most important Web technologies that commonly find application in a medical framework. We start with a brief overview of the utilization of simple Web sites and the World Wide Web and continue to discuss XML data format, mobile agents, Web Services, and security related issues.

6.2   THE WORLD WIDE WEB

Health related sites in the World Wide Web have reached an impressive population. It is estimated that as many as 2% of all Web sites focus on health-related information. The degree to which these resources are used is similarly high, with an estimated 50–75% of Internet users having looked online for such information at least once [4, 5]. Clearly, rather than such a wealth of uncontrolled information, what is needed by individuals and clinicians alike is clear, accurate, and well-presented information that one may rely on [6]. In this direction a number of attempts have been made to filter the reliable from the not so reliable medical information portals [7, 8].

Of course, the application of Web in e-Health is not limited to health-related portals; much more sophisticated applications have been developed, which use the Web as their supporting technology. As simple examples, [9] presents a Web-based tool for nutrition counselling of patients at high cardiovascular risk, while [10] also combines the e-mail service, and [11] develops customized browser plug-ins to suit both the expert and the novice Internet user patient. Decision support, more details on which are presented in Chapter 5, is also found in a Web-based setting [12, 13].

Asynchronous communication is another area in which the Web provides a solution to networked e-Health; as clinicians cannot be available online 100% of the time, communication can be facilitated by tools that allow for asynchronous exchange of messages and medical data. There are today a number of Web sites that have been developed specifically for this purpose. Through a store and forward approach, asynchronous communication can be achieved between patient and doctor [14]; this finds application mainly in the case of remote and chronic patients who are not closely and directly supervised by a clinician. The asynchronous option is not only supported for the communication between patients and doctors, but also for the communication and interaction between different doctors [15].

A situation when networked e-Health proves its worth the most is disaster. As disasters come across with no warning and at any place and time, it is practically impossible to have the required resources, this including experienced medical personnel, in place in order to cope with them; this is a gap that Web-based medical applications have been proven able to cover, in both military and civilian circumstances. Medical networks set up specifically to provide the infrastructure needed to support networked e-Health in a disaster framework include Global Health Network (GHNet) [16], Global Health Disaster Network (GHDNet) [17], Relief Web [18], and Virtual Information Center (VIC) [19], while the Rapid Health Assessment Module also provides for quick sentence translation so that disaster relief can be provided in a multilingual setting [20].

6.3   XML DATA REPRESENTATION

Similarly to any other system increasing in scale, networked e-Health systems are often formed from the integration of multiple parts, often ones designed and developed by different people, in the framework of different projects and sometimes even at different times. In order for the integration of such components to be feasible, and the compatibility of future components

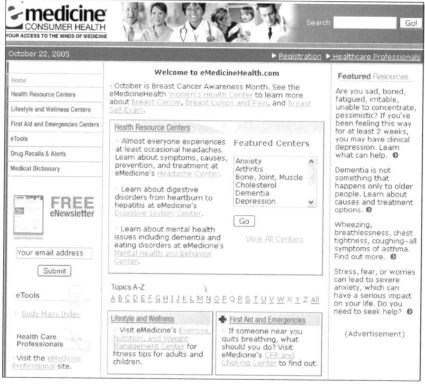

FIGURE 6.1: http://www.emedicinehealth.com/ portal

to be assured, data representation issues need to be addressed in a formal manner. Ideally, single representation standard would exist and all applications would adhere to it, making the integration of any e-Health components a much simpler task.

As seen in Chapter 6, a number of standards have been proposed in the medical field, and as a result useful medical information is already coded in a number of different formats in digital information systems around the world. Certainly, telemedicine and e-Health applications can benefit from the consideration of this information; rather than moving toward a totally novel standard, the research community is focusing on ways to facilitate the exchange of data stored in these existing formats over different types of Internet connections.

The problem of uniform data exchange over the Internet has been addressed by W3C with the extensible mark-up language (XML) standard [21]. Using XML Schema (which has rendered the use of Document Type Definitions obsolete) one may define the structure and permitted contents of an XML file. Once the XML Schema has been shared, applications developed independently by different people at different times and using different software development platforms can safely exchange data using any networking channel; safely, of course,

```xml
<?xml version="1.0" encoding="ISO-8859-1" ?>
<HL7_Message>
  <MSH>
    <Field_Separator>.</Field_Separator>
    <Encoding_Characters>ASCI</Encoding_Characters>
    <Sending_Application>HL7 Message Creator</Sending_Application>
  </MSH>
  <PID>
    <Patient_ID>001</Patient_ID>
    <Patient_Name>Ilias Maglogiannis</Patient_Name>
    <Maiden_Name />
    <Date_of_Birth>1/1/1970</Date_of_Birth>
    <Sex>Male;</Sex>
    <Phone_Number>22730-82239</Phone_Number>
    <Patient_Address>University of Aegean Karlovasi Samos</Patient_Address>
    <Nationality>Hellenic</Nationality>
    <Marital_Status>Single;</Marital_Status>
  </PID>
  <PV1>
    <Current_Date>2005-10-20 15:34:40</Current_Date>
    <Medical_Supervisor>Dr. Manolis Wallace</Medical_Supervisor>
    <Institution>Athens General Hospital</Institution>
    <Institution_Address />
    <Exam_Type>Blood and Urin Test</Exam_Type>
    <Comments>Sent to LIS</Comments>
  </PV1>
</HL7_Message>
```

FIGURE 6.2: An example of an HL7 XML document

referring to the certainty of their ability to correctly connect to each other rather than to the security of any established connection.

It is already clear that XML is the way to go for networked e-Health applications too. Efforts have been made to combine XML with existing medical standards such as Health Level Seven (HL7) [22] and Digital Imaging and Communications in Medicine (DICOM) [23], some new XML-based standards have been proposed [24], while XML in general is being used in numerous European and international systems to facilitate data exchange between end-users often located in different countries.

6.4   WEB SERVICES AND MOBILE AGENTS

It is not only data format that one needs to consider on the Internet; the way to establish communication is an issue itself at times. In this section we discuss Web services and mobile agents, two technologies that aim to address issues related to connectivity and the way to achieve it.

Web servers using HTTP to format pages and display information offer for a very intuitive data presentation format that users can easily relate to. On the other hand, without strong natural language processing and understanding tools, it is not possible for software components to automatically extract useful information from Web sites. Similarly, although search engines offer for very intuitive searches by providing both simple and complex input options, their proprietary form makes it impossible to develop generic software modules to perform searches using them.

In order to allow for the Web present information and services to become available to software components in an automated manner, W3C has proposed using Web Services to share services and the XML-based Web services description language (WSDL) to describe them [21]. Similarly to XML and XML Schema, once the query and result parameters for a Web Service have been described using WSDL and the file has been shared, any informed application is able to successfully query the service and parse the provided results.

In Fig. 6.3 we present a typical setup for a Web service-based application. Web Services formalize the structure of the exchanged data, so that server and client components can be developed independently and in possibly quite different platforms, while the SOAP protocol formalized the format of the data, thus making the application independent of the underlying data communication channel.

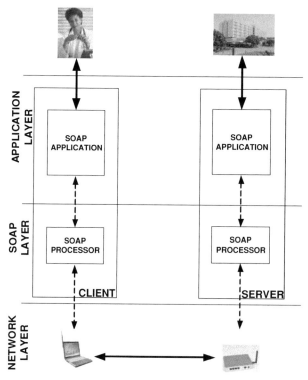

FIGURE 6.3: A sample architecture for Web Service application

Although Web Service technology is quite new, its excellent characteristics have allowed it to have an early adoption in various fields, networked e-Health being one of them. More and more medical systems are developed using Web Services and WSDL to describe the developed features, rather than simply using XML and sharing XML Schemas [25–28].

Other issues one might choose to consider include that of services that might be unavailable at a given time, as well as that of large amounts of data needing to be processed for a simple conclusion to be reached. In both cases the traditional client–server approach proves to be suboptimal, with a lot of time being spent online before the required information is acquired. As an example of online time being too expensive to maintain for long, one can see, for example, the case of Antarctica. There, with INMARSAT having to be used in order to support for exchange of medical data, a different approach needs to be followed [29].

Mobile agents have been proposed in order to tackle such problems. They are agents that can physically travel across a network and perform tasks on machines that provide agent-hosting capability. This allows processes to migrate from computer to computer, for processes to split into multiple instances that execute on different machines, and to return to their point of origin. Their benefits include:

- *Task delegation at offline time.* Once the agent has been sent, the local host can be disconnected from the Internet and only connected at a later time again in order to receive the agent together with its findings.
- *Bandwidth conservation.* When data and network intensive queries need to be performed in order to reach a conclusion or to acquire statistics, the agent travels to the host of the database, runs the queries locally, and returns with the results.
- *Privacy.* With agents examining information locally, sensitive data do not have to be transmitted over the Internet any more.
- *Parallel processing.* Agents also possess the ability to split themselves, so that different copies are sent at the same time to different locations.

Mobile agents have also found application in the medical field. For example, the following networked e-Health systems utilize agent technology [29–31].

6.5   SECURITY ISSUES

Security-related issues have played a major role as obstacles to the further penetration of the Internet and its services in everyday life. For example, *Business Week* has provided evidence in the past that the main user reservation for using the Internet is the lack of privacy, rather than cost-related issues. Obviously, when it comes to the exchange of sensitive private information, especial medical information, user reservation is even higher. Additionally, the legal issues that need to be considered are equally severe, causing service providers, i.e. medical organizations such as hospitals, to also have reservations [32]. As the contribution made by telemedicine and e-Health is evident and immense, this is not an issue of whether the Internet and communications in general should be used in a medical framework. The question is rather how these communications shall be protected from the different types of risks that lurk in network environments.

First of all one needs to identify the risks, starting from what it is that may be in danger. In the case of networked e-Health, there are two main risks to consider: security and privacy violations. Although often mistakenly treated as synonyms, the two terms refer to distinct and equally important properties. Security of the data refers to the protection of their content, while privacy refers to the protection of the identity of their source or owner. For these to be protected, technical, procedural, and organizational measures need to be taken by organizations and/or individuals communicating sensitive data over insecure channels [33]. As far as the technical aspect is concerned, it is typically general purpose Information Security Technologies (ISTs) and Privacy-Enhancing Technologies (PETs) that are being ported in the medical framework in order to enhance protection levels.

An important effort is being made in order to standardize the way to pursue protection at an international level [34], and works are presented that attempt to set the guidelines for secure and private communications in the medical sector through the careful examination of real life scenarios [33]. Unfortunately, a common conclusion is that regardless of the technical solutions available, major attention needs to be given to procedural and organizational issues, as systems' security is often jeopardized by user actions far before malicious attacks of a technical nature take place.

When it comes to the specific technologies and methodologies used in order to maximize privacy and security, with the exception of some standardization bodies, such as CEN and HL7, that have proposed Electronic Healthcare Record architectures that address issues of authorization, access permissions, and data exchange between different applications [35], there are no networked e-Health specific tools; it is the general purpose security techniques that are applied in the medical field as well. And although there have been at times concerns that augmented security might imply reduced utility [36], it is now generally accepted that security measures can be transparent and have no obvious effect on the amount and quality of services available.

Given that communication channels are insecure by nature, the most common technical measure used to ensure security is that of encrypted data exchange. Using the Secure Sockets Layer (SSL) protocol, data can travel over insecure channels and third parties are able to intercept it, but only the intended recipient is able to decrypt the message in order to retrieve the original data. The process relies on the public key infrastructure (PKI) [37, 38].

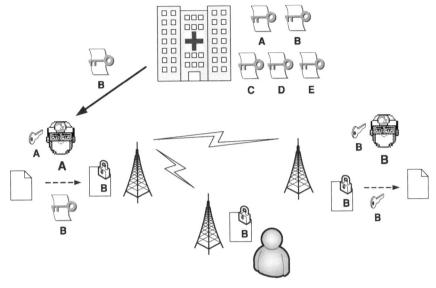

FIGURE 6.4: Using the PKI for data exchange

A key is a sequence of symbols which is utilized by the SSL algorithms in order to process data and produce an encrypted form of it. The complexity of the algorithm and the length of the key are such that a brute force attack on an encrypted message is computationally intractable, and thus encrypted data are generally considered as secure. The process that generates the public key also produces as a combined product the private key. This is another sequence of symbols of equal length that can be utilized by a similar decrypting algorithm in order to retrieve the original message. The first key is made public by publishing in the PKI, so that everyone can send secure messages to a given recipient, while the second is kept private at the recipient, and thus the names public and private key.

In Fig. 6.4 we outline the process of secure communication utilizing public and private keys. The PKI authority issues public/private key pairs for every authorized host and makes the public ones freely available to anyone that requests them. In order to securely transmit a document to host B, host A requests the public key that corresponds to B and utilizes it to encrypt the message. Thus, even if a third part intercepts the communication, the intercepted document is encrypted and cannot be accessed without the private key that only host B has.

Although this assures to a high degree the security of the transmitted data, it leaves another door open for fraud: third parties may be unable to intercept transmitted data, but they are able to impersonate the other end of the communication, thus intercepting sensitive data stored on the other end. In order to protect from this possibility, a reliable method for identity verification is required. IP address checks, MAC address checks, username password sessions, and even using encrypted transmission and frequent challenges as in the Challenge Handshake

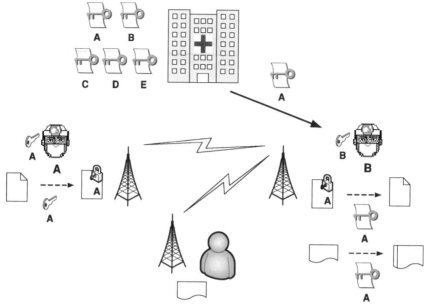

FIGURE 6.5: Using the PKI for identity verification

Authentication Protocol (CHAP) are all imperfect measures that do not guarantee absolute security.

The most reliable known approach to identity authentication is again provided by the PKI and relies on the commutative property of the keys: alternatively to using the public key for encryption and the private one for encryption, the private one may be used on the encrypting end and the public one on the decrypting end, as seen in Fig. 6.5. This process does not guarantee data security, as any third party is able to use the publicly available public key to perform the decryption. What it does provide for is reliable identity verification as only the real sender possesses the private key and is able to correctly encrypt the data; data transmitted by any third party cannot be meaningfully decrypted by the recipient, and thus are rejected as generated by an unauthorized sender.

In real life it is customary to maximize security by utilizing the sender's private key in order to reencrypt the message, thus providing for identity verification, and then using the recipient's public key to encrypt the data, thus providing for data security. At the recipient end the reverse process is followed, with the recipient's private key used first and the sender's public key second.

Although the SSL and the PKI provide a reliable solution for secure communications over the Internet, when wireless *ad hoc* networks are utilized encryption can only provide for reduced security levels. The reason behind this is mainly that portable networking devices typically have reduced computational capabilities and are unable to process adequately complex keys and algorithms when real-time exchange of data is required. Thus, specialized (and typically reduced in performance) encryption methodologies need to be applied [39, 40]. Actually, security concerns and corresponding solutions exist for any type of connectivity pursued, as well as for any type of protocol or methodology utilized; one can consider security in Web servers, in XML documents, in mobile agents, and so on.

What one typically finds in real life installations is the combination of different security measures, each one focusing at different levels and stages of the communication process, thus protecting from different types of vulnerabilities [32, 41].

REFERENCES

[1] G. Kormentzas, I. Maglogiannis, D. Vasis, D. Vergados, and A. Rouskas, "A modeling and simulation framework for compound medical applications in regional healthcare networks," *Int. J. Electron. Healthc.*, vol. 1, no. 4, pp. 427–441, 2005. doi:10.1504/IJEH.2005.006689

[2] M. S. Cashen, P. Dykes, and B. Gerber, "eHealth technology and Internet resources: barriers for vulnerable populations," *J. Cardiovas. Nurs.*, vol. 19, no. 3, pp. 209–214, 2004.

[3] G. R. Eysenbach, "Towards ethical guidelines for e-health: JMIR theme issue on ehealth ethics," *J. Med. Internet Res.*, vol. 2, no. 1, p. e7, 2000.

[4] H. Tatsumi, H. Mitani, Y. Haruki, and Y. Ogushi, "Internet medical usage in Japan: current situation and issues," *J. Med. Internet Res.*, vol. 3, no. 1, p. e12, 2001. doi:10.2196/jmir.3.1.e12

[5] M. Murero, G. D'ancona, and H. Karamanoukian, "Use of the Internet by patients before and after cardiac surgery: telephone survey," *J. Med. Internet Res.*, vol. 3, no. 3, p. e27, 2001.doi:10.2196/jmir.3.3.e27

[6] J. Powell and A. Clarke, "The WWW of the World Wide Web: who, what, and why?" *J. Med. Internet Res.*, vol. 4, no. 1, p. e4, 2002.doi:10.2196/jmir.4.1.e4

[7] S. Lapinsky and S. Mehta, "Web reports: critically appraising online resources," *Crit. Care*, vol. 6, no. 6, p. 462, 2002.doi:10.1186/cc1830

[8] F. R. Jelovsek, "Assessing the credibility of medical web sites," OBG Manage., vol. 12, no. 9, 2000.

[9] M. Verheijden, P. van Genuchten, M. Godwin, C. van Weel, and W. van Staveren, "Heartweb: a web-based tool for tailoring nutrition counselling in patients at elevated cardiovascular risk," *Technol. Health Care*, vol. 10, no. 6, pp. 555–557, 2002.

[10] J. Castrén, "Ask doctor per email—a modern health service among Finnish university students, carried out by general practitioners," *Technol. Health Care*, vol. 10, no. 6, pp. 458–459, 2002.

[11] C. Lau, R. S. Churchill, J. Kim, F. A. Matsen, and Y. Kim, "Asynchronous web-based patient-centered home telemedicine system," *IEEE Trans. Biomed. Eng.*, vol. 49, no. 12, pp. 1452–1462, 2002.doi:10.1109/TBME.2002.805456

[12] D. Jegelevicius, V. Marozas, A. Lukosevicius, and M. Patasius, "Web-based health services and clinical decision support," *Stud. Health Technol. Inform.*, vol. 105, pp. 27–37, 2004.

[13] G. Eysenbach, J. Powell, M. Englesakis, C. S. Rizo, and A. Stern, "Health related virtual communities and electronic support groups: systematic review of the effects of online peer to peer interactions," *Br. Med. J.*, vol. 328, no. 7449, pp. 1–6, 2004.

[14] H. M. Kim, J. C. Lowery, J. B. Hamill, and E. G. Wilkins, "Accuracy of a web-based system for monitoring chronic wounds," *Telemed. J. e-Health*, vol. 9, no. 2, pp. 129–140, 2003. doi:10.1089/153056203766437471

[15] C. E. Chronaki, F. Chiarugi, E. Mavrogiannaki, C. Demou, P. Lelis, D. Trypakis, M. Spanakis, M. Tsiknakis, and S. C. Orphanoudakis, "An eHealth platform for instant interaction among health professionals," *Comput. Cardiol.*, vol. 30, pp. 101–104, 2003. doi:full_text

[16] F. Sauer, R. E. LaPorte, S. Akazawa, E. Boostrom, E. Ferguson, C. Gamboa, *et al.*, "Global health network task force. Towards a global health network," *Curr. Issues Public Health*, vol. 1, pp. 160–164, 1995.

[17] I. M. Libman, R. E. LaPorte, S. Akazawa, E. Boostrom, C. Glosser, E. Marler, E. Pretto, F. Sauer, A. Villasenor, F. Young, and G. Ochi, "The need for a global health disaster network," *Prehospital Disaster Med.*, vol. 12, no. 1, pp. 11–12, 1997.

[18] http://www.reliefweb.int/

[19] R. E. LaPorte, "Improving public health via the information superhighway," *The Scientist*, vol. 11, p. 10, 1997.

[20] Dragon Systems, *Rapid Health Assessment Module for Disaster Management and Humanitarian Assistance: Operations Guide*, Newton, MA: Dragon Systems, 1997.

[21] http://www.w3.org/

[22] M. Tobman, C. Nätscher, H. Sussmann, and A. Horsch, "A new approach for integration of telemedicine applications into existing information systems in healthcare," *Stud. Health Technol. Inform.*, vol. 90, pp. 152–155, 2002.

[23] A. Anagnostaki, S. Pavlopoulos, E. Kyriakou, and D. Koutsouris, "A novel codification scheme based on the 'VITAL' and 'DICOM' standards for telemedicine applications Anagnostaki," *IEEE Trans. Biomed. Eng.*, vol. 49, no. 12, pp. 1399–1411, 2002. doi:10.1109/TBME.2002.805458

[24] A. Anagnostaki, S. Pavlopoulos, and D. Koutsouris, "XML and the VITAL standard: the document-oriented approach for open telemedicine applications," *MEDINFO*, vol. 10, no. 1, pp. 77–81, 2001.

[25]   V. Bicer, G. Laleci, A. Dogac, and Y. Kabak, "Artemis message exchange framework: se-
mantic interoperability of exchanged messages in the healthcare domain," *ACM SIGMOD
Rec.*, vol. 34, no. 3, pp. 71–76, 2005.doi:10.1145/1084805.1084819

[26]   R. J. Rodrigues and A. Risk, "ehealth in Latin America and the Caribbean: development
and policy issues," *J. Med. Internet Res.*, vol. 5, no. 1, p. e4, 2003.doi:10.2196/jmir.5.1.e4

[27]   G. Demiris, "Disease management and the Internet," *J. Med. Internet Res.*, vol. 6, no. 3,
p. e33, 2004.doi:10.2196/jmir.6.3.e33

[28]   J. Anhøj and L. Nielsen, "Quantitative and qualitative usage data of an Internet-based
asthma monitoring tool," *J. Med. Internet Res.*, vol. 6, no. 3, p. e23, 2004.

[29]   S. Pillon and A. R. Todini, "eHealth in Antarctica: a model ready to be transferred to
every-day life," *Int. J. Circumpolar Health*, vol. 63, no. 4, pp. 436–442, 2004.

[30]   S. Kirn, C. Heine, R. Herrler, and K.-H. Krempels, "Agent.hospital—a framework
for clinical applications," in *Applications of Software Agent Technology in the Health Care
Domain*, A. Moreno and J. Nealon, Eds. Basel: Birkhäuser Publisher, 2003, pp. 67–85.

[31]   M. Rodriguez, J. Favela, V. M. Gonzalez, and M. Munoz, "Agent based mobile collaboration and information access in a healthcare environment," *e-Health: Appl. Comput. Sci.
Med. Health Care*, vol. 5, pp. 133–148, 2003.

[32]   B. Stanberry, *The Legal and Ethical Aspects of Telemedicine*. London: Royal Society of
Medicine Press, 1998.

[33]   A. Gritzalis, C. Lambrinoudakis, D. Lekkas, and S. Defteros, "Technical guidelines for enhancing privacy and data protection in modern electronic medical environments," *IEEE Trans. Inf. Technol. Biomed.*, vol. 9, no. 3, pp. 413–423, 2005.
doi:10.1109/TITB.2005.847498

[34]   P. Ruotsalainen and H. Pohjonen, "European security framework for healthcare," *Stud.
Health Technol. Informat.*, vol. 96, pp. 128–134, 2003.

[35]   KONA Editorial Group, "Patient record architecture, HL7 Document (Online),"
http://www.hl7.org/special/committees/sgml/PRA/PRA_ballot_April00.zip, 2000.

[36]   N. P. Terry, "An ehealth diptych: the impact of privacy regulation on medical error and
malpractice litigation," *Am. J. Law Med.*, vol. 27, no. 4, pp. 361–419, 2001.

[37]   Sax *et al.*, "Wireless technology infrastructures for authentication of patients: PKI that
rings," *J. Am. Med. Informat. Assoc.*, vol. 12, pp. 263–268, 2005.

[38]   G. Kambourakis, I. Maglogiannis, and A. Rouskas, "PKI-Based Secure Mobile Access
to Electronic Health Services and Data," *Technol. Healthc.*, vol. 13, no. 6, pp. 511–526,
2005.

[39]   P. K. Dutta "Security considerations in wireless sensor networks," in *Sensors Expo*, San
Jose, CA, 2004.

[40]   F. Hu and N. K. Sharma, "Security considerations in wireless sensor networks," *Ad hoc Netw.*, vol. 3, no. 1, pp. 69–89, 2005.doi:10.1016/j.adhoc.2003.09.009

[41]   V. Traver, C. Fernández, E. Montón, J. C. Naranjo, S. Guillén, and B. Valdivieso, "IST project Integrated Distributed Environment for Application Services in e-Health (IDeAS)," Jan. 31, 2006, http://www.ideas-ehealth.upv.es.

CHAPTER 7

Distributed Collaborative Platforms for Medical Diagnosis over the Internet

7.1 INTRODUCTION

Hospitals are nowadays sufficiently rich in their infrastructure to handle the internal administrative and clinical processes for their inpatients. However, this kind of information exchange is usually reduced at regional level, and the need to integrate the processes in geographically distributed and organizationally independent organizations and medical personnel has arisen. This challenge leads the design of health information applications to combine the principles of heterogeneous Workflow Management Systems (WfMSs), the peer-to-peer networking architecture, the World Wide Web, and the Visual Integration concepts to facilitate medical information exchange among geographically distributed personnel.

Under this scope, the goal of this chapter is to present the concept of Web-based collaborative systems for the facilitation of physician's communication at remote locations. Such systems enable communication via audiovisual means through a public network. The Health Level Seven (HL7), the Clinical Data Architecture (CDA), and the Digital Imaging and Communications in Medicine (DICOM) standards [1, 2] are used for the transfer and storing of the medical data.

7.2 STATE OF THE ART REVIEW

The development of a distributed collaborative system has received the attention of several research groups. For instance, Collaboratory for Research on Electronic Work (CREW) [8] is a typical example of a collaborative system implemented in the University of Michigan. The specific computer-based environment is able to give multiple access to shared documents and provides CREW community with information about other CREW members, their activities, location, availability, and interests. Moreover, it supports collaboration tools such as postable notes and e-mail.

Healthcare information collaborative systems preoccupy the medical informatics research community as well. InterMed [9] is a system aiming at the development of a robust framework for medical collaboration, using the Internet as the basic network. The project TeleMed [10] supports both real-time and off-line sessions. Multiple physicians at remote locations are able to view, edit, and annotate patient data on real-time. This project focuses on security and confidentiality services, access to online knowledge, clinical data access and sharing, as well as collaborative policies, procedures, and guidelines. The ARTEMIS project aims at "advance cooperative activities of healthcare providers to promote the delivery of total and real-time care" [11]. A group of developers, physicians, and healthcare researchers are using prototypes and commercial off-the-shelf technologies to develop an open collaboration healthcare environment. In ARTEMIS, community care networks consist of a collection of primary and specialized care providers, who collaborate to meet the healthcare needs of the community. The communication infrastructure allows primary care physicians to consult with remote specialists with computer support for X-rays, ultrasound, voice-annotations, and other multimedia information.

Another project focusing on the healthcare collaboration domain is the virtual medical office [12], which is an integrated environment that encourages users to take active participation in the management of their health. It provides access to digital medical libraries, yellow pages, and regional healthcare resources.

WebOnCOLL [13] has been designed in the context of the regional healthcare network of Crete and employs the infrastructure of regional healthcare networks to provide integrated services for virtual workspaces, annotations, e-mail, and on-line collaboration. Virtual workspaces support collaborative concepts like personal Web pages, bulletin boards, discussion lists, shared workspaces, and medical case folders. Annotations provide a natural way for people to interact with multimedia content, while e-mail is one of the most popular forms of communication today. Finally, another type of health collaborative application is described in [14]. The presented application uses a multiuser environment for collaboration with multimodal human machine interaction. Audio–video teleconferencing is also included to help the radiologists to communicate with each other simultaneously while they are working on the mammograms.

7.3   COLLABORATIVE ARCHITECTURAL ASPECTS

Examining networked applications from an architectural aspect, we find that the most common architectures used today are the client–server, the 3-tier or n-tier, and the peer-to-peer networking schemas. Client–Server scheme is the "network architecture in which each computer or process on the network is either a *client* or a *server*. Servers are powerful computers or processes dedicated to managing disk drives (*file servers*), printers (*print servers*), or network traffic (*network servers*). Clients are PCs or workstations on which users run applications. Clients rely on servers for resources, such as files, devices, and even processing power." N-tier application architecture [7] provides a model for developers to create a flexible and reusable application. By breaking up an application into tiers, developers only have to modify or add a specific layer, rather than having to rewrite the entire application over, if they decide to change technologies or scale up. In the term "N-tier," "N" implies any number—like 2-tier or 4-tier, basically any number of distinct tiers used in a specific architecture. Peer to Peer [15, 16] on the other hand is the "type of network in which each workstation has equivalent capabilities and responsibilities. This differs from client–server architectures, in which some computers are dedicated to serving the others."

The upside of the peer-to-peer architecture is that it is more dependable, reliable, and available, as there is no single point of failure; instead information is usually replicated at several peers. In addition it is relatively inexpensive, it causes low network workload by skipping extra reroutes, and it is fairly simple to set up and manage. Furthermore, it scales better because there is no single point of access since there is no machine called server, and also no server farm. The flip side is that it is limited in extensibility, tends to overburden user workstations by having them play the role of server to other users, is largely insecure, and is typically unable to provide system-wide services since the typical workstation will run a standard desktop operating system incapable of hosting any major service (e.g., a post office). Table 7.1 summarizes the benefits and the fault backs of the peer-to-peer and the client–server architectures related to the needs of the medical collaboration platform [15, 16].

7.4   COLLABORATIVE PLATFORM—MAIN FUNCTIONS AND MODULES

The subsections below describe the main functions and modules available in most of medical collaborative platforms.

7.4.1   Users' Registration

Each physician wishing to join a collaboration session has to register him- or herself at a medical collaborative platform. During the registration procedure, the physician provides the system with information about his/her identity and specialty. In most cases the registration is a face-to-face process in order to ensure the integrity of the physician's personal information.

7.4.2   On-line Calls

In order to initiate a session, a physician (named physician A) selects from the "on-line medical personnel" list the remote physician (physician B) whom he/she wishes to contact. His/her module sends a communication request to the selected doctor. On the remote doctor's side, physician B receives the call and accepts it. Physician B's module returns to the initiator of the communication an acknowledgment. Such a procedure is depicted in the sequence diagram of Fig. 7.1.

TABLE 7.1: Client–Server vs Peer-to-Peer Networking Architectures

PLATFORM REQUIREMENTS	PEER-TO-PEER	CLIENT–SERVER
Collaboration sessions: Best quality of communication	Point-to-point communication, direct exchange of data, no extra reroutes,	Extra rerouting through the server module
Centralized physician management	Does not support centralized control	Centralized user and network management
Security schema	Distributed security schema, difficult to set up and manage	Centralized security schema, easy to control
Administrative control	No need of a central administrator, limited responsibilities	The existence of an administrator is crucial, various and essential responsibilities
Use of a centralized medical database	No centralized control over the database	Server can provide centralized control of database access list, easy data distribution
Expandability	Significantly easy to expand, low complexity, ad hoc infrastructure	High complexity on expanding architecture, Slow-moving on new adaptations
Availability	High level of availability due to distributed architecture	Low level of availability in case of traffic or server malfunction, i.e., system crash, denial of service

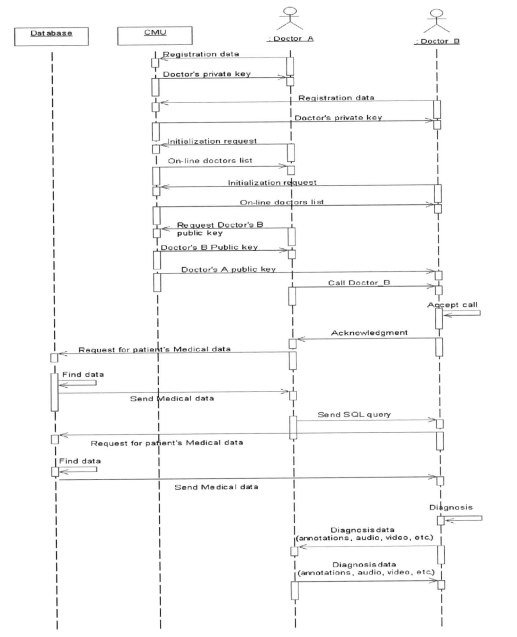

FIGURE 7.1: Communication sequence diagram of a collaborative session

Assuming that physician A wishes to retrieve medical record data of a specific patient, he/she fills up the related form with the patient's name and the specific data and commits a query to a middleware entity (called hereafter Collaboration Management Unit, CMU). Upon receiving the data, physician A informs physician-B device about the patient's name and the committed SQL query. Physician B from his/her part follows a similar procedure and receives the patient's data as well. As soon as both physicians successfully retrieve the requested part of the patient's record, they can manipulate the data and exchange opinion about patient's condition.

7.4.3   Patient Record and Medical Image Transfer

Patient's medical data are stored in an electronic health record database that connects with the CMU middleware database. Most of computer-based patient systems follow the standards for incident-based electronic health record (EHR) storage, including information such as demographic data, medical history, and previous diagnoses and medical examinations (such as medical images, ECG/EEG biosignals, etc). Medical images comply with the DICOM format [4]; therefore, a DICOM viewer is incorporated in such platforms.

7.4.4   Video–Audio Communication

Most of the medical collaborative platforms enable video and audio conference as well as message communication. On the lowest level of the applications, standards are responsible for translating, sending, and receiving application's information. Information can be transferred over the Internet or corporate intranet using both TCP and UDP connections. TCP is used primarily for data transport and call control, while UDP enables secondary connections for transmitting and receiving audio and video. Winsock provides the interface to the network stack and maps information between the programs and the network.

Most of the collaborative applications support H.323 for audio and video conferencing. The ITU H.323 standard for audio and video conferencing provides a set of standard audio and video codecs capable of encoding and decoding the input and output from audio and video sources that is transmitted over various connection rates and networks that do not provide a guaranteed quality of service. By using H.323, the application can support many different modes of Internet telephony and can send and receive audio and video information to H.323-compatible nodes. H.323 computers and equipment can carry real-time video, audio, and data, or any combination of these elements. This standard is based on the Internet Engineering Task Force (IETF) Real-Time Protocol (RTP) and Real-Time Control Protocol (RTCP), with additional protocols for framing, call signaling, call control, data and audiovisual communications, and for negotiating the information flow and sequencing.

7.4.5   Workspace Management

The workspace of the doctor's module consists of the basic image-processing tools offering the capability of interaction in real time with his/her colleague located on the other end of the session. Upon loading the medical images on the DICOM viewer, all editing taking place on his/her side is automatically displayed on the image of the other party. Note that one part at a time can have access to the editing tools.

The features of the tools mostly implemented are the following:

- *Image comments.* Doctors can comment the image by writing particular observations on specific points of interest.

- *Use of annotations.* This feature enables user to draw geometrical schemes with his/her mouse selection ROIs (region of interests) and relay this information to the other end.

- *Multiple color channels* Physicians are able to select different colors or spectrum channels for image viewing.

- *Contrast and brightness configuration.* Medical personnel can set via scrollbars the contrast and the brightness level of the medical image displayed.

- *Image scaling (zoom-in, zoom-out).* Medical personnel can increase and decrease the zoom level of a picture via special buttons.

7.4.6   Image Compression

The most amount of bandwidth during the on-line collaboration sessions is consumed by medical images. The data throughput is significantly enhanced if the images are compressed. Many compression algorithms have been presented in the past. These can be classified to lossless and lossy. In the former case the reconstructed image is identical to the uncompressed one while in the latter there is loss of information. All the reviewed platforms have the capability of transmitting lossless compressed images.

7.4.7   Security Issues

Specific attention in medical collaborative platforms is drawn toward the security requirements. The transmission of sensitive medical data related to the patient heath record through a public network arises the problem of communication privacy. Most of the approaches followed include the incorporation of encryption algorithms that achieve confidentiality, authenticity, and integrity of the data and the adoption of PKI infrastructure for consistent user authorization and authentication. The middleware CMU entity applies and manages the system's security policy. The policy is predetermined by the system administrator and authorized by the health organization that controls the system. The policy consists of specific rules assigning physicians to authorization groups. The group definition is based on the geographical regions; i.e., the group

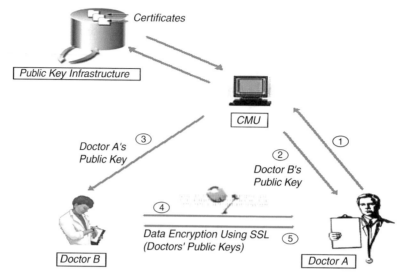

FIGURE 7.2: A PKI-based security scheme

of North Aegean consists of physicians whose hospitals are placed in the North Aegean region, group of South Aegean, etc. Each group has specific rights over patients' data; i.e., a doctor that belongs to the North Aegean group can access medical data stored on hospital databases placed in that specific region. Since patients' data are stored on databases, specific countermeasures have to be taken to ensure data confidentiality. In order to meet this requirement each group receives a username and password that provide read access to databases.

Note that during collaboration both physicians need to have access to data related to the patient cross-examined. What happens in case that only one part of the communication has access to that information? One possible solution is to allow the authorized doctor to forward the information directly to the nonauthorized doctor. However, this solution would be inefficient in terms of performance, since it would cause additional load to doctor's device. A better approach would include a daemon running on CMU machine dedicated to users' access management. The basic idea is to create a temporary user account for nonauthorized users with a specific validity period. By using this account users can have access to the patient's data during a specific session. As soon as the session ends the temporary account is removed from the database. A security scheme based on public key infrastructure (PKI) is depicted in Fig. 7.2.

7.5   CONCLUSIONS AND CHALLENGES FOR THE FUTURE

The chapter discusses the domain specific technological issues regarding medical collaboration platforms. Considering the fact that physicians are not computer experts, it is of high importance to build a user-friendly platform, adopting most of the desired functionality that users need during a collaboration session. Taking into account the great amount of data that a medical record has, the platform requires medium bandwidth resources; however, the data throughput may be enhanced by compression methods. Remaining future challenges are the enhancement of the main functionalities in terms of speed, user friendliness, HCI, ergonomics, data compression and security, and the transition to the mobile paradigm by implementing platforms to work on personal digital assistants (PDAs) and smart phones.

REFERENCES

[1] *Health Level Seven, Reference Information Model.* Ann Arbor, MI: Health Level Seven, Inc., 2002. http://www.hl7.org.

[2] *Digital Imaging and Communications in Medicine (DICOM),Part 1: Introduction and Overview.* Rosslyn, VA: National Electrical Manufacturers Association, 2001.

[3] http://www.rsna.org/practice/dicom/intro/index.html

[4] The ACR-NEMA DICOM Standard, http://medical.nema.org/

[5] J. F. Coyle, A. R. Mori, and S. M. Huff, "Standards for detailed clinical models as the basis for medical data exchange and decision support," *Int. J. Med. Inf.*, vol. 69, pp. 157–174, 2003.doi:10.1016/S1386-5056(02)00103-X

[6] T. Penzel, B. Kemp, G. Klösch, A. Schlögl, and J. Hasan, "Acquisition of Biomedical Signals Databases," *IEEE Eng. Med. Biol.*, pp. 25–32, May–Jun. 2001. doi:10.1109/51.932721

[7] http://www.15seconds.com/issue/011023.htm

[8] A. Arbor, "Collaboratory for Research on Electronic Work," http://www.crew.umich.edu/CREW.home.html.

[9] InterMed Collaboratory, http://smi-web.stanford.edu/projects/intermedweb/Overview.html.

[10] D. G. Kilman and D. W. Forslund, "An international collaboratory based on virtual patient records," *Commun. ACM*, vol. 40, pp. 111–117, Aug. 1997. doi:10.1145/257874.257898

[11] V. Jagannathan *et al.*, "An overview of the CERC Artemis project," Concurrent Eng. Res. Center, West Virginia Univ., Morgantown, Tech. Rep. CERC-TR-RN-95-002, Apr. 1995, http://www.cerc.wvu.edu/nlm/artemis/summary.html.

[12] Virtual telemedicine Office, http://www.telemedical.com

[13] C. E. Chronaki, D. G. Katehakis, X. C. Zabulis, M. Tsiknakis, and S. C. Orphanoudakis, "WebOnCOLL: medical collaboration in regional healthcare networks," *IEEE Trans. Inf. Technol. Biomed.*, vol. 1, no. 4, pp. 257–269, Dec. 1997.doi:10.1109/4233.681170

[14] M. Akay, I. Marsic, A. Medl, and G. Bu, "A system for medical consultation and education using multimodal human/machine communication," *IEEE Trans. Inf. Technol. Biomed.*, vol. 2, no. 4, pp. 282–291, Dec. 1998.doi:10.1109/4233.737584

[15] G. Cugola and G. P. Picco, "Peer-to-Peer for collaborative applications," in *Proc. Int. Workshop Mobile Teamwork Support, co-located with the 22nd International Conference on Distributed Computing Systems*, Vienna, H. Gall and G. P. Picco, Eds., Piscataway, NJ: IEEE Press, Jul. 2002, pp. 359–364.

[16] K. Albrecht, R. Arnold, and R. Wattenhofer, "Clippee: a large-scale client/peer system," in *Int. Workshop Large-Scale Group Communication, held in conjunction with the 22nd Symposium on Reliable Distributed Systems (SRDS)*, Florence, Italy, October 2003.

[17] S. Saha, "Image compression—from DCT to wavelets: a review," *ACM Crossroads*, http://www.acm.org/crossroads/2000.

[18] A. Cohen, I. Daubechies, and J. C. Feauveau, "Biorthogonal bases of compactly supported wavelets," *Comm. Pure Appl. Math.*, vol. 45, pp. 485–560, 1992.

[19] A. Shamir and E. Tromer, "On the cost of factoring RSA-1024," *RSA CryptoBytes*, vol. 6, no. 2, pp. 10–19, 2003.

[20] A. K. Lenstra and E. R. Verheul, "Selecting cryptographic key sizes," *J. Cryptol.*, vol. 14, no. 4, pp. 255–293, 2001.

[21] J. M. Shapiro, "Embedded image coding using zerotrees of wavelet coefficients," *IEEE Trans. Signal Process.*, vol. 41, no. 12, pp. 3445–3462, Dec. 1993.doi:10.1109/78.258085

[22] A. Said and W. A. Pearlman, "A new fast and efficient image codec based on set partitioning in hierarchical trees," *IEEE Trans. Circuits Syst. Video Technol.*, pp. 243–250, Jun. 1996.doi:10.1109/76.499834

[23] A. Järvi, J. Lehtinen, and O. Nevalainen, "Variable quality image compression system based on SPIHT," *Signal Process.: Image Commun.*, 1998.

[24] J. Lehtinen, "Limiting distortion of a wavelet image codec," *Acta Cybern.*, vol. 14, no. 2, pp. 341–356, 1999.

[25] M. Vettereli and J. Kovačević, *Wavelets and Subband Coding*. Englewood Cliffs, NJ: Prentice Hall, 1995.

[26] W. Pennebaker and J. Mitchell. *JPEG: Still Image Data Compression Standard*. New York: Van Nostrand Reinhold, 1992.

CHAPTER 8

Distributed Telemedicine Based on Wireless and *Ad Hoc* Networking

8.1 INTRODUCTION

The introduction of small and portable personal digital assistants (PDAs), mobile phones, and wireless communication, in general, has already changed the way we communicate with each other for everyday routine interaction. In the healthcare sector this reality enabled the development of advanced wireless distributed telemedicine services for remote areas.

In recent years several telemedicine applications have been successfully implemented over wired communication technologies like POTS (Plain Old Telephone Services), ISDN, and xDSL. However, nowadays, modern wireless telecommunication means like the GSM, the GPRS, and the UMTS mobile telephony standards, as well as satellite communications and Wireless Local Area Network (WLAN) allow the operation of wireless telemedicine systems freeing the medical personnel and/or the subject-monitored bounded to fixed locations [1].

Telemedicine applications, including those based on wireless technologies, span the areas of emergency healthcare, telecardiology, teleradiology, telepathology, teledermatology, teleophtlalmology, and telepsychiatry. In addition, health telematics applications enabling the availability of prompt and expert medical care have been exploited for the provision of healthcare services at understaffed areas like rural health centers, ambulance vehicles, ships, trains, airplanes, as well as for home monitoring [1].

The scope of the present chapter is twofold. It provides, in its first part, a concise report in the enabling wireless technologies that are applied in the distributed and mobile telemedicine sector [2]. The second part of the paper presents the special features for digital signals, images, and videos transmission in mobile telemedicine systems and discusses the application of wireless *ad hoc* networks in telemedicine. A wireless *ad hoc* network is a collection of two or more wireless devices that have the possibility of wireless communication and networking without the help of any infrastructure. The chapter presents the particular characteristics of communication with *ad hoc* networks in mobile telemedicine applications and provides examples of such *ad hoc* networking utilization.

8.2   EMERGING MOBILE TELEMEDICINE TECHNOLOGIES

In this section we highlight the main wireless communications technologies that have been used and are about to be used in mobile telemedicine systems, namely GSM/GPRS, satellite, 3G (UMTS), DVB-H/T (Digital Video Broadcasting for Television/Handheld devices), WPAN (Wireless Personal Area Network), and WLAN (Wireless Personal Area Network) [1]. These systems are summarized in Table 8.1.

GSM is a system currently in use and is the second generation (2G) of the mobile communication networks. When it is in the standard mode of operation, it provides data transfer speeds of up to 9.6 kb/s. Through the years the HSCSD technique makes it possible to increase the data transmission to 14.4 kb/s or even triple at 43.3 kb/s [3, 4].

The evolution of mobile telecommunication systems from 2G to 2.5G (iDEN 64 kb/s, GPRS 171 kb/s, EDGE 384 kb/s) systems provides the possibility of faster data transfer rates, thus enabling the development of telemedicine systems that require high data transfer rates and are currently only feasible on wired communication networks [3].

The fast development of current 3G wireless communication and mobile network technologies will be the driving force for future developments in mobile telemedicine systems. 3G wireless technology represents the convergence of various 2G wireless systems. One of the most important aspects of 3G technology is its ability to unify existing cellular standards under one umbrella [3].

In recent years other mobile network technologies such as WLANs and WPANs have become popular. WLAN allows users to access a data network at high speeds of up to 54 Mb/s as long as users are located within a relatively short range (typically 30–50 m indoors and 100–500 m outdoors) of a WLAN base station (or antenna). In the US WLAN operates in two unlicensed bands: a) 802.11b and 802.11g operate in the 2.4 GHz band, together with many other devices including Bluetooth and cordless telephones and b) 802.11a (Wi-Fi 5.2 GHz) operates in the 5.2 GHz band, which at this point is relatively free of interference from other electrical devices operating in this band [5, 6].

WPANs are defined with IEEE standard 802.15. They are considered quite important for the e-health sector, and though they are included in Table 8.1 they will be discussed separately in the next chapter, which presents pervasive computing in e-health.

The recent surge research of mobile *ad hoc* networking will also trigger a parallel research activity in the application of these emerging technologies for mobile telemedicine. It is expected that 4G will integrate existing wireless technologies including UMTS, GSM, WLAN, WPAN, and other newly developed wireless technologies into a seamless system. Some expected key features of 4G networks are stated as: a) high usability b) support for telemedicine services at low transmission cost, c) 4G provides personalized services in order to meet the demands of different users for different services, and d) 4G systems provide facilities for integrating services.

TABLE 8.1: Main Wireless Communication Networks/Standards

TYPE	SUBTYPE	FREQUENCY BAND	DATA TRANSFER RATES	CORRESPONDING TELEMEDICINE APPLICATIONS
Wireless LAN/PAN	IEEE 802.11a	5 GHz	54 Mb/s	• Access to medical data in hospital environments • In hospital patient telemonitoring • Home care and smart homes applications
	IEEE 802.11b	2.4 GHz	11 Mb/s	
	IEEE 802.11g	2.4 GHz	54 Mb/s	
	HIPERLAN 1	5 GHz	23.5 Mb/s	
	HIPERLAN 2	5 GHz	54 Mb/s	
	Bluetooth	2.4 GHz	1 Mb/s	
	HomeRF 1.0	2.4 GHz	2 Mb/s	
	HomeRF 2.0	2.4 GHz	10 Mb/s	
GSM	GSM-900	900 MHz	9.6–43.3 kb/s	• Emergency telemedicine • Patient telemonitoring in random locations
	GSM-1800	1800 MHz	9.6–43.3 kb/s	
	GSM-1900	1900 MHz	9.6–43.3 kb/s	
GPRS	GPRS	900/1800/ 1900 MHz	171.2 kb/s	
3G/UMTS	UMTS		2 Mb/s	• High bandwidth mobile telemedicine
Satellite	ICO	C, S band	2.4 kb/s	• Location based telemedicine • High bandwidth telemedicine • Regional healthcare networking infrastructure
	Globalstar	L, S, C band	7.2 kb/s	
	Iridium	L, Ka band	2.4 kb/s	
	Cyberstar	Ku, Ka band	400 kb/s–30 Mb/s	
	Celestri	Ka band, 40–50 GHz	155 Mb/s	
	Teledesic	Ka band	16 kb/s–64 Mb/s	
	Skybridge	Ku band	16 kb/s–2 Mb/s	
DVB	DVB-T	5/6/7/8 MHz	5–32 Mb/s	
	DVB-H	5/6/7/8 MHz	10 Mb/s	

The main technological characteristics of 4G systems are expected to be transmission speeds higher than in 3G (min 50–100 Mb/s, average 200 Mb/s), system capacity larger than in 3G by ten times, transmission costs per bit 1/10 to 1/100 of that of 3G, support for Internet protocols (IPv6), various quality of service (QoS) providing many kinds of best effort multimedia services corresponding to users demand, and user-friendly services where users can access many services in a short time span as compared with other wireless systems of longer waiting times for response.

4G advances will provide both mobile patients and citizens the choices that will fit their lifestyle and make easier for them to interactively get the medical attention and advice they need; when and where it is required and how they want it regardless of any geographical barriers or mobility constraints. The concept of including high-speed data and other services integrated with voice services is emerging as one of the main points of future telecommunication and multimedia priorities with the relevant benefits to citizen-centered healthcare systems [3]. These creative methodologies will support the development of new and effective medical care delivery systems into the 21st Century. The new wireless technologies will allow both physicians and patients to roam freely, while maintaining access to critical patient.

Finally, DVB-T/H stands for digital video broadcasting for television/handheld devices, and it is the standard for digital, terrestrial television in Europe and beyond. The MPEG-2 technology for data compression and multiplexing makes it possible to use the existing frequency channels for a multiplied number of TV programs or equivalents. Furthermore, IP data can be encapsulated directly into the MPEG-2 transport stream using the multiprotocol encapsulation method (MPE), standardized within the DVB system. By this means, a DVB-T or DVB-H network can be designed for mobile reception in order to deliver medical data and IP packet streams [10].

8.3   DISTRIBUTED MOBILE TELEMEDICINE REQUIREMENTS

The range and complexity of telecommunications technology requirements vary with specific medical or health applications. However, generically defined digital medical devices impose the telecommunications performance requirements. Table 8.2 illustrates a sampling of several of the more common digital medical devices that may be used in distributed telemedicine. Except for the last few items contained in the table (starting with ultrasounds and running through full motion video), the majority of vital sign medical devices require relatively low data transmission rates [11].

Regarding the transmission of medical images, there are essentially no theoretical bandwidth requirements, but lack of bandwidth needs longer transmission time. Yet, high-quality medical images such as a single chest radiograph may require from 40 to 50 Megabytes. In

TABLE 8.2: Data Rates of Typical Devices Used in Telemedicine

DIGITAL DEVICE	IMAGE RESOLUTION		DATA RATE REQUIRED
	SPATIAL	CONTRAST	
Digital blood pressure monitor (sphygmomanometer)	−	−	<10 kb/s
Digital thermometer	−	−	< 10 kb/s
Digital audio stethoscope and integrated electrocardiogram	−	−	<10 kb/s
Ultrasound, cardiology, radiology	512 × 512	×8	256 KB (image size)
Magnetic resonance image	512 × 512	×8	384 KB (image size)
Scanned x-ray	1024 × 1250	×12	1.8 MB (image size)
Digital radiography	2048 × 2048	×12	6 MB (image size)
Mammogram	4096 × 4096	×12	24 MB (image size)
Compressed and full motion video (e.g., ophthalmoscope, proctoscope, etc.)	−	−	384 kb/s to 1.544 Mb/s (speed)

practice, it is desirable to transmit medical images during a single patient visit, so as to at least avoid a follow-up visit.

Medical image compression techniques have primarily focused on lossless methods, where the image has to be reconstructed exactly from its compressed format due to the diagnostic use. The medical compression issue has been thoroughly discussed in Chapter 4.

About the digital video compression, the DICOM committee has not yet adopted any standard. The adoption of MPEG-2 is possible, but this is limited by the MPEG-2 requirement for constant delay method for frame synchronization. On the other hand, the transmission of off-line video is still possible. It is important to distinguish among the requirements for real-time video transmission, off-line video transmission, medical video and audio for diagnostic applications, and nondiagnostic video and audio. Real-time video transmission for diagnostic applications is clearly the most demanding. Off-line video transmission is essentially limited by the requirement to provide patient–doctor interaction. Real-time diagnostic audio applications include the transmission of stethoscope audio or the transmission of the audio stream that accompanies the diagnostic video. Good quality diagnostic audio at 38–128 kb/s using Dolby AC-2 has been achieved, while MPEG-1 Layer 2 audio (32–256 kb/s) or Dolby AC-3 (96–768 kb/s) may also be used. For nondiagnostic applications such as teleconferencing, H.261 (64 kpbs–1.92 Mb/s) and H.263 (15–64 kb/s) may be acceptable. A typical application will require a diagnostic audio and video bitstream, in addition to a standard teleconferencing bitstream [1, 13, 14].

Reviewing literature, from 150 papers (including both conference and journal papers) published under the mobile telemedicine from year 1979 until now, we selected 50 distributed applications. These systems covered a significant component of the whole spectrum of health telematics applications grouped under the wireless technologies of GSM/GPRS, satellite, radio, and WLAN and into the areas of emergency healthcare, telecardiology, teleradiology, telepsychology, teleopthalmology, and remote monitoring. The majority of these applications utilize the GSM network and the areas of emergency telemedicine, remote monitoring, and telecardiology. Satellite links were also used in many telemedicine applications and most of them were used for remote monitoring. Satellite systems have the advantage of worldwide coverage and offer a variety of data transfer speeds, even though satellite links have the disadvantage of high-operating cost. Finally, WLAN technology is an emerging technology already applied for emergency telematics and teleradiology [1, 13–17].

8.4   *AD HOC* NETWORKS IN DISTRIBUTED MOBILE TELEMEDICINE APPLICATIONS

Mobile *ad hoc* networks represent complex distributed systems that consist of wireless mobile nodes, which can freely and dynamically self-organize into arbitrary and temporary, "*ad hoc*" network topologies, allowing people and devices to seamlessly internetwork in areas with no pre-existing communication infrastructure or centralized administration. In such an environment, provision of medical data communication is currently dependent on low-bandwidth point-to-point telemedicine systems. An easily field-deployable *ad hoc* wireless network or even a basic sensor network could provide a significant improvement in real-time patient monitoring and emergency communications capability in general [18, 19].

From the standards that have already presented for transmission in mobile telemedicine networks, the ones that have the possibility to support *ad hoc* communication or could do it through modifications and changes are: a) IEEE 802.11, b) HIPERLAN/2, c) Bluetooth, and d) HomeRF [6, 7, 20]. The Bluetooth and HomeRF standards could not be compared with IEEE 802.11 and HIPERLAN/2 because their possibilities are more for a WPAN than for a WLAN.

In a prehospital environment such as a patient's home it is important to capture his/her clinical data and transfer them to a diagnostic center (DC) for earlier diagnosis, with reasonable speed. A mobile *ad hoc* network with various devices such as PDAs, laptops, digital electrocardiogram machines, and cellular phones may be established in this case.

The advancement of mobile *ad hoc* network technology has the potential to allow a physician to get patient's information anywhere and even before the patient reaches the hospital. Timely access of patient's information may fundamentally improve patient care in both pre- and inhospital setting with doctors' earlier access of patient information and thus intervention. For example in prehospital environment, patient's data such as the electrocardiogram of a patient with chest pains, and other clinical data, are rarely send to hospital before the ambulance arrival [21].

In addition, inhospital environment the doctors and the nurses want sometimes to communicate based on not constant infrastructure. Similarly, in hospital the visiting doctors and the stretcher-bearers could communicate with the internal doctors and receive or transmit information through *ad hoc* networks [18, 22].

Another case that *ad hoc* networks use in mobile telemedicine systems is Wireless Sensor Networks (WSNs). WSNs consist of a large number of tiny devices capable of executing sensing, data processing, and communication tasks and as such can be applied in numerous user scenarios in healthcare monitoring. WSNs consist of a subcategory of WPANs, and they will be presented in the next chapter.

REFERENCES

[1] C. S. Pattichis, E. Kyriacou, S. Voskarides, M. S. Pattichis, R. Istepanian, and C. N. Schizas, "Wireless telemedicine systems: an overview," *IEEE Antennas Propag. Mag.*, vol. 44, no. 2, pp. 143–153, 2002.doi:10.1109/MAP.2002.1003651

[2] K. Shimizu, "Telemedicine by mobile communication," *IEEE Eng. Med. Biol. Mag.*, vol. 18, no. 4, pp. 32–44, Jul.–Aug. 1999.doi:10.1109/51.775487

[3] R. S. H. Istepanian, E. Jovanov, and Y. T. Zhang, "Introduction to the special section on m-health: beyond seamless mobility and global wireless health-care connectivity (Guest editorial)," *IEEE Trans. Inf. Technol. Biomed.*, vol. 8, no. 4, pp. 405–414, 2004. doi:10.1109/TITB.2004.840019

[4] B. Woodward and M. F. A. Rasid, "Wireless telemedicine: the next step?" in *Proc. 4th Annu. IEEE Conf. Information Technology Applications in Biomedicine*, Birmingham, UK, 2003, pp. 43–46.doi:full_text

[5] K. Banitsas, R. S. H. Istepanian, and S. Tachkara, "Applications of medical wireless LAN (MedLAN) systems," *Int. J. Health Mark.*, vol. 2, no. 2, pp. 136–142, 2002.

[6] B. P. Crow, I. Widjaja, J. G. Kim, and P. T. Sakai, "IEEE 802.11 wireless local area networks," *IEEE Commun. Mag.*, vol. 35, no. 9, pp. 116–126, Sept. 1997.doi:10.1109/35.620533

[7] J. C. Haartsen, Ericsson Radio Systems B.V. "The bluetooth radio system," *IEEE Pers. Commun.*, vol. 7, no. 1, pp. 28–36, Feb. 2000.doi:10.1109/98.824570

[8] A. M. Hura, "Bluetooth—enabled teleradiology: applications and complications," *J. Digit. Imaging*, vol. 15, suppl. 1, pp. 221–223, 2002.doi:10.1007/s10278-002-5067-0

[9] ZigBee Alliance, "IEEE 802.15.4, ZigBee standard," http://www.zigbee.org

[10] DVB—Digital Video Broadcasting, http://www.dvb.org

[11] M. Ackerman, R. Craft, F. Ferrante, M. Kratz, S. Mandil, and H. Sapci, "Telemedicine technology," *Telemed. J. e-Health*, vol. 8, no. 1, pp. 71–78, 2002. doi:10.1089/15305620252933419

[12] NEMA—DICOM Standards Committee, (2004). "DICOM Strategic Document, Version 4.0, December 29 2004 http://medical.nema.org/dicom/geninfo/Strategy.htm

[13] International Telecommunication Union, Telecommunication Development Bureau, ITU-D Study Group 2, "Question 14-1/2: application of telecommunications in health care, technical information—a mobile medical image transmission system in Japan, PocketMIMAS," Rapporteurs Meeting Q14-1/2 Japan, 2004.

[14] C. LeRouge, M. J. Garfield, and A. R. Hevner, "Quality attributes in telemedicine video conferencing," in *Proc. 35th Hawaii Int. Conf. System Sciences*, 2002, p. 159.

[15] S. Tachakra, X. H. Wang, R. S. H. Istepanian, and Y. H. Song, "Mobile e-Health: the unwired evolution of telemedicine," *Telemed. J. e-Health*, vol. 9, no. 3, pp. 247–257, 2003. doi:10.1089/153056203322502632

[16] U. Engelmann, A. Schröter, E. Borälv, T. Schweitzer, and H.-P. Meinzer, "Mobile teleradiology: all images everywhere," *Int. Cong. Series 1230*, pp. 844–850, 2001. doi:10.1016/S0531-5131(01)00143-1

[17] E Kyriacou, S Pavlopoulos, A Berler, M Neophytou, A Bourka, A Georgoulas, A Anagnostaki, D Karayiannis, C Schizas, C Pattichis, A Andreou, and D Koutsouris, "Multi-purpose healthcare telemedicine systems with mobile communication link support," *Biomed. Eng. Online*, vol. 2, p. 7, 2003.doi:10.1186/1475-925X-2-7

[18] C.-K. Toh, *Ad Hoc Mobile Wireless Networks—Protocols and Systems*. New Jersey: Prentice Hall PTR, pp. 243–255, 266–283, 2002.

[19] R. Ramanathan and J. Redi, "A brief overview of ad hoc networks: challenges and directions," *IEEE Commun. Mag.*, vol. 40, 50th Anniversary Commemorative Issue, pp. 20–22, 2002.doi:10.1109/MCOM.2002.1006968

[20] J. Khun-Jush, G. Malmgren, P. Schramm, and J. Torsner, "HIPERLAN type 2 for broadband wireless communication," *Ericson Review*, vol. 77, no. 2, pp. 108–119, 2000.

[21] V. Anantharaman and L. Swee Han, "Hospital and emergency ambulance link: using IT to enhance emergency pre-hospital care," *Int. J. Med. Inf.*, vol. 61, no. 2–3, pp. 147–161, 2001.doi:10.1016/S1386-5056(01)00137-X

[22] E. Jovanov, A. O'Donnell Lords, D. Raskovic, P. G. Cox, R. Adhami, and F. Andrasik, "Stress monitoring using a distributed wireless intelligent sensor system," *IEEE Eng. Med. Biol. Mag.*, vol. 22, no. 3, pp. 49–55, May–Jun. 2003.doi:10.1109/MEMB.2003.1213626

CHAPTER 9

Ambient Intelligence and Pervasive Computing for Distributed Networked E-Health Applications

9.1 INTRODUCTION

In this chapter we present the concepts of ambient intelligence and pervasive computing in electronic healthcare systems. Both concepts refer to the availability of software applications and medical information anywhere and anytime and the invisibility of computing; computing modules are hidden in multimedia information appliances, which are used in everyday life [1]. Pervasive healthcare systems refer mostly in patient remote monitoring or telemonitoring, which is an important sector of distributed telemedicine. Patient telemonitoring involves the sensing of a patient's physiological and physical parameters and transmitting them to a remote location, typically a medical center, where expert medical knowledge resides [11, 18].

A typical telemonitoring system has the ability to record physiological parameters and provide information to the doctor in real time through a wireless connection, while it requires sensors to measure parameters like arterial blood pressure, heart rate, electrocardiogram (ECG), skin temperature, respiration, and glucose or patient position and activity [19]–[23]. Filtered signals and medical data are either stored locally on a monitoring wearable device for later transmission or directly transmitted, e.g., over the public phone network, to a medical center. Such architecture is depicted in Fig. 9.1.

The vital signals that may be monitored by pervasive system and corresponding dependant variables are presented in Table 9.1. Each biosignal provides different and complementary information on a patient's status, and for each specific person the anticipated range of signal parameters is different. For example, heart rate may vary between 30 and 250 beats/min for normal people in different circumstances; likewise, breathing rate could be between 5 and 50 breaths/min. electroencephalogram (EEG) and ECG are considerably more complex biosignals with spectra up to 10 KHz. The voltage levels of the collected biosignals also vary from microvolts (EEG) to millivolts (ECG). The biosignals' characteristics are summed up in Table 9.2.

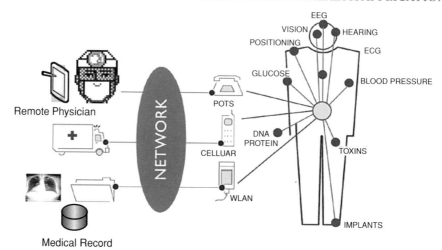

FIGURE 9.1: A typical pervasive healthcare system architecture

9.2 ENABLING TECHNOLOGIES IN PERVASIVE HEALTHCARE

The development and the standardization of the mobile IEEE 802.x family protocols [7], along with the, currently available, 2.5G and 3G networks and the positioning technologies provide sufficient tools for the development of such "intelligent medical environments" realizing location aware pervasive e-health services.

9.2.1 Networking Technologies

Regarding networking, there are two main enabling technologies according to their topology: *on-body* (wearable) and *off-body* networks. Recent technological advances have made possible a new generation of small, powerful, and mobile computing devices. A wearable computer must be small and light enough to fit inside clothing. Occasionally, it is attached to a belt or

TABLE 9.1: Physiological Signals and Dependent Variables

SIGNAL	DEPENDANT VARIABLES
ECG	Heart rate, ventricular beat, ST segment, QT time
Noninvasive blood pressure	Systolic, diastolic, mean, pulse rate
Respiration	Breath rate, expired CO_2
Pulse oximetry	Pulse rate, pulse volume, oxygen saturation

TABLE 9.2: Biosignals Characteristics

BIOMEDICAL MEASUREMENTS	VOLTAGE RANGE (V)	NUMBER OF USERS = K (SENSORS)	BANDWIDTH (HZ)	SAMPLE RATE (HZ)	RESOLUTION (B/SAMPLE)	INFORMATION RATE (B/S)
ECG	0.5–4 m	5–9	0.01–250	1250	12	15 000
Heart sound	Extremely small	2–4	5–2000	10 000	12	120 000
Heart rate	0.5–4 m	2	0.4–5	25	24	600
EEG	2–200 μ	20	0.5–70	350	12	4200
EMG	0.1–5 m	2+	0–10 000	50 000	12	600 000
Respiratory rate	Small	1	0.1–10	50	16	800
Temperature of body	0–100 m	1+	0–1	5	16	80

other accessory or is worn directly like a watch or glasses. An important factor in wearable computing systems is how the various independent devices interconnect and share data. An off-body network connects to other systems that the user does not wear or carry, and it is based on a wireless local area network (WLAN) infrastructure, while an on-body or wireless personal area network (WPAN) connects the devices themselves—the computers, peripherals, sensors, and other subsystems—and runs at *ad hoc* mode. Tables 9.3 and 9.4 present the characteristics of off-body and on-body networking technologies respectively.

WPANs are defined within the IEEE 802.15 standard. The most relevant protocols for pervasive e-health systems are Bluetooth and ZigBee (IEEE 802.15.4 standard). Bluetooth technology was originally proposed by Ericsson in 1994, as an alternative to cables that linked mobile phone accessories. It is a wireless technology that enables any electrical device to communicate in the 2.5-GHz ISM (license free) frequency band [6]. It allows devices such as mobile phones, headsets, personal digital assistants (PDAs) and portable computers to communicate and send data to each other without the need for wires or cables to link the devices together. It has been specifically designed as a low-cost, low-size, and low-power radio technology, which is particularly suited to the short range of a personal area network (PAN). The main features of Bluetooth are a) real-time data transfer usually possible between 10–100 m, b) support of point-to-point wireless connections without cables, as well as point-to-multipoint connections to enable *ad hoc* local wireless networks, and c) data speed of 400 kb/s symmetrically or 700–150 kb/s of data asymmetrically. On the other hand, ZigBee (IEEE 802.15.4 standard) has been developed as a low data rate solution with multimonth to multiyear battery life and very low complexity. It is intended to operate in an unlicensed international frequency band. The maximum data rates for each band are 250, 40, and 20 kb/s. The 2.4-GHz band operates worldwide while the sub-1-GHz band operates in North America, Europe, and Australia.

Pervasive healthcare systems set high demanding requirements regarding energy, size, cost, mobility, connectivity, and coverage. Varying size and cost constraints directly result in corresponding varying limits on the energy available, as well as on computing, storage, and communication resources. Low power requirements are necessary also from safety considerations since such systems run near or inside the body. Mobility is another major issue for pervasive e-health applications because of the nature of users and applications and the easiness of the connectivity to other available wireless networks. Both off-body and PANs must not have line-of-sight (LoS) requirements.

9.2.2   Positioning Technologies

Pervasive services often require specific infrastructure for the estimation of the user's location. Several techniques that provide such estimation are currently available. The most prominent are satellite-based (e.g., GPS) or terrestrial infrastructure-based (e.g., Cell-ID, TOA).

TABLE 9.3: Off-Body Networking Technologies [25]

TECHNOLOGY	OPERATIONAL SPECTRUM	PHYSICAL LAYER	CHANNEL ACCESS	MAXIMUM DATA RATE	COVERAGE
Wireless LANs					
IEEE 802.11b	2.4 GHz	DSSS with PBCC	CSMA-CA	22 Mb/s	100 m
HomeRF	2.4 GHz	FHSS with FSK	CSMA-CA and TDMA	10 Mb/s	>50 m
HiperLAN2	5 GHz	OFDM with QAM	Central resource control/ TDMA/TDD	32–54 Mb/s	30–150 m
Cellular telephony					
CDMA20 00 1 × Ev-DO	Any existing frequency; band with 2/1 × 3.75 MHz channels	DSSS with QPSK/BPSK; 1.2288 MHz or 3.6864 MHz	FDD or TDD	2 Mb/s	Area of a cell
UMTS (WCDMA)	1920–1980 MHz and 2110–2170 MHz; 2 × 5 MHz channels	DSSS with QPSK; 3.84 MHz (chip rate)	FDD	2.048 Mb/s	Area of a cell
ReFLEX (two-way paging)	150–900 MHz; 25- or 50-KHz channel	FSK	TDMA-based	9600 b/s in, 6400 b/s out	Typically a large city

(Continued)

TABLE 9.3: Off-Body Networking Technologies [25] (*Continued*)

TECHNOLOGY	OPERATIONAL SPECTRUM	PHYSICAL LAYER	CHANNEL ACCESS	MAXIMUM DATA RATE	COVERAGE
Wireless MANs					
Ricochet	900 MHz	FHSS	CSMA	176 Kb/s	As of Sept. 2003, only Denver and San Diego, USA
IEEE 802.16	10–66 GHz; line of sight required; 20/25/28 MHz channels	TDMA-based; uplink QPSK; 16- and 64-QAM	FDD and TDD variants	120/134.4 Mb/s for 25/28 MHz channel	Typically a large city

TABLE 9.4: On-Body Networking Technologies [25]

TECHNOLOGY	OPERATIONAL SPECTRUM	PHYSICAL LAYER	CHANNEL ACCESS	MAXIMUM DATA RATE	COVERAGE
IrDA	Infrared; 850 nm	Optical rays	Polling	4 Mb/s	<10 m
MIT/IBM personal area networks	0.1–1 MHz	DSSS with ASK/FSK	CSMA-CA/CD, TDMA, or CDMA	417 Kb/s (theoretical); 2400 b/s (IBM); 9600 b/s (MIT)	Typically the entire human body
BodyLAN	Not available	OOC (time-based spread spectrum)	Complex TDMA scheme	32 Kb/s	<10 m
Fabric area networks	125 KHz; RFID	RF fields	Polling	1–10 Kb/s (duplex); 100 Kb/s (simplex)	Multiple short-range antennas (<2 cm each)
IEEE 802.15.1 (Bluetooth)	2.4 GHz ISM band	FHSS;	Master-slave polling, TDD	Up to 1 Mb/s	<10 m
IEEE 802.15.3	2.402–2.480 GHz ISM band	Uncoded QPSK, trellis-coded QPSK, or 16/32/64-QAM scheme	CSMA-CA and GTS in superframe structure	11–55 Mb/s	<10 m
IEEE 802.15.4	2.4 GHz and 868/915 MHz	DSSS with BPSK or MSK (O-QPSK)	CSMA-CA and GTS in superframe structure	868 MHz– 20 Kb/s, 915 MHz–40 Kb/s, and 2.4 GHz– 250 Kb/s	<20 m

Satellite-based positioning does not operate properly in deep canyons and indoors where cellular coverage may be denser. Terrestrial-based positioning may be more imprecise with sparse deployment of base stations in rural environments, where satellite visibility is better. Position-fixing in indoor environments may also exploit other technologies such as WLAN. WLAN-based positioning requires access points (AP) to be installed on the building (e.g., hospital) structure and WLAN-equipped portable devices. In general, two types of WLAN positioning architectures exist. In the *centralized scheme* a server calculates the position of a client based, e.g., on triangulation/trilateration or scene analysis. Example of the centralized architecture is the Ekahau's Positioning Engine [3], which provides absolute information (i.e., depicted on a floor plan, building grid, or on a geographic coordinates system) and works for 802.11 and Bluetooth. The *distributed scheme* relies on specialized software that runs on clients (e.g., PDAs). Each client uses its own radio frequency (RF) measurements in order to calculate its position. Examples of such architecture are the Microsoft RADAR [1] and the Nibble [2] systems. RADAR uses both scene analysis and triangulation through the received signal's attenuation. It has been developed for 802.11 networks. It provides a positioning accuracy of 3–4 m, but with a precision of 50%. It uses the scene analysis technique to estimate location and provides symbolic (the position of a located object is determined in reference to the observer or another known object) and absolute positioning information. The accuracy of the system can reach 3 m. Other proprietary radio- or infrared-based positioning schemes also exist for indoor environments. Indicative examples of such systems are the Active Badge and Active Bat systems. An extensive survey of similar platforms can be found in [4].

9.3   PERVASIVE HEALTHCARE APPLICATIONS IN CONTROLLED ENVIRONMENTS

The use of pervasive systems in controlled environments, such as hospitals, may be divided into two broad categories. The first one relates to applications, enabling the mobile ubiquitous delivery of medical data and implementations of mobile electronic health records (EHR), accessible by PDA's or Tablet PC's in a hospital equipped with WLAN infrastructure [13]. Several research groups [11, 14, 15, 17] have experimented on the use of handheld computers of low cost and high portability, integrated through a wireless local computer network within the IEEE 802.11 or Bluetooth standards. Regarding the medical data exchange, DICOM (www.dicom.org) and HL7 (www.hl7.org) standards are used in the data coding and transmission via mobile client/server applications capable of managing health information.

On the other hand, pervasive systems are used for the monitoring and diagnosis of patients. A wide range of medical monitors and sensors may enable the mobile monitoring of a patient, who is able to walk freely without being tied to the bed. Pervasive systems in hospital environment are mostly based on Bluetooth communication technology. For example,

Khoor *et al.* [10] have used the Bluetooth system for short-distance (10–20 m) data transmission of digitized ECGs together with relevant clinical data. Hall *et al.* have demonstrated a Bluetooth-based platform for delivering critical health record information in emergency situations [11], while J. Andreasson *et al.* have developed a remote system for patient monitoring using Bluetooth enabled sensors [12]. The above examples show that the merging of mobile communications and the introduction of handhelds along with their associated technology has potential to make a big impact in emergency medicine. Moreover, many market projections indicate that mobile computer is both an emerging and enabling technology in healthcare [13].

9.4 PERVASIVE HEALTHCARE IN DISTRIBUTED NONHOSPITAL SETTINGS—HOME CARE APPLICATIONS

Facilities for medical practice in nonhospital settings are limited by the availability of medical devices suitable for producing biosignals and other medical data. There are a number of active research and commercial projects developing sensors and devices, which do not require local intervention to enable contact with a clinician remote from the care environment. These new systems provide automated connection with remote access and seamless transmission of biological and other data upon request. Pervasive systems in nonhospital systems aim at better managing of chronic care patients, controlling of health delivery costs, increasing quality of life and quality of health services, and the provision of distinct possibility of predicting and thus avoiding serious complications [8].

The patient or elder person mainly requires monitoring of his/her vital signals (i.e., ECG, blood pressure, heart rate, breath rate, oxygen saturation, and perspiration). Patients recently discharged from hospital after some form of intervention, for instance, after a cardiac incident, cardiac surgery, or a diabetic comma are less secure and require enhanced care. The most common forms of special home monitoring are ECG arrhythmia monitoring, post surgical monitoring, respiratory and blood oxygen levels monitoring, and sleep apnoea monitoring. In the case of diabetics, the monitoring of blood sugar levels resigns the patient to repeated blood sampling, which is undesirable and invasive. One possible solution is the development of implantable wireless sensor devices that would be able to give this information quickly, and in a continuous fashion. Current conditions where home monitoring might be provided include hypertension, diabetes (monitoring glucose), obesity and CHF (monitoring weight), asthma and COPD (monitoring spirometry/peak flow), and, in the near future, conditions utilizing oximetry monitoring. Other home-monitoring conditions might include preeclampsia, anorexia, low-birth-weight infants, growth abnormalities, and arrhythmias. Most chronic health conditions in children and adults could be managed and/or enhanced by home monitoring.

In most applications two monitoring modes are foreseen: the Batch and the Emergency Mode. Batch mode refers to the every day monitoring process, where vital signs are acquired and transmitted periodically to a health-monitoring center. The received data are monitored by the doctor on-duty and then stored into the patient's electronic health record maintained by a healthcare center. The emergency mode occurs because the patient does not feel well and, thus, decides to initiate an out-of-schedule session, or because the monitoring device detects a problem and automatically initiates the transfer of data to the corresponding center. Application of emergency episode detection and the corresponding alarm processing are important for the protection of the patient. An alarm represents a change in the status of a physiological condition or a sensor reading state outside of agreed limits.

9.5   CONCLUSIONS AND FUTURE CHALLENGES

The technological advances of the last few years in mobile communications, location-aware and context-aware computing have enabled the introduction of pervasive healthcare applications. These applications may assist medical personnel and people with health-related problems in two distinct ways. One concerns the remote monitoring of patients with chronic diseases, involving diagnosis and monitoring using biosignal sensors and the patients' history record, and the immediate notification of a doctor or a medical center in the case of an emergency. The other is toward dealing with the everyday practice in hospital settings, involving the monitoring of the inpatient's status and the mobile access to medical data.

However, the use of such pervasive healthcare raises several challenges. Personal data security and location privacy are considered to be the most important. Furthermore, pervasive healthcare systems are very critical systems, as they deal with a person's health, and therefore they raise high standards regarding reliability, scalability, privacy-enhancing, interoperability, and configurability, among other things. On the other hand, since healthcare systems are intended to be used by low or medium computer literate users, usability issues come to the foreground.

Regardless of the remaining challenges, pervasive healthcare systems are anticipated to be expanded in the near future by using the most recent technological advances in a more active and direct way for offering more comprehensive and higher quality health services to citizens.

REFERENCES

[1]   P. Bahl and V. N. Padmanabhan, "RADAR: An in-building RF-based user location and tracking system," in *Proc. IEEE INFOCOM*, Tel Aviv, Israel in March 2000, pp. 775–784.

[2]   P. Castro, P. Chiu, T. Kremenek, and R. Muntz, "A probabilistic room location service for wireless networked environments," in *Proc. Ubicomp 2001, LNCS 2201*, G. D. Abowd, B. Brumitt, and S. A. N. Shafer, Eds. Springer-Verlag, 2001.

[3]   Electronic information available at http://www.ekahau.com

[4]   J. Hightower and G. Borriello, "Location systems for ubiquitous computing," *IEEE Computer*, vol. 34, no. 8, pp. 57–66, Aug. 2001 (also in IEEE special report "Trends Technologies and Applications in Mobile Computing").

[5]   G. Abowd, "Software engineering issues for ubiquitous computing," in *Proc. Int. Conf. Software Engineering*, 1999, pp. 5–84.

[6]   Bluetooth Consortium, http://www.bluetooth.com

[7]   http://standards.ieee.org/

[8]   N. Maglaveras *et al.*, "Home care delivery through the mobile telecommunications platform: The citizen health system (CHS) perspective," *Int. J. Med. Inf.*, vol. 68, pp. 99–111, 2002.

[9]   J. Birnbaum, "Pervasive information systems," *Commun. ACM*, vol. 40, no. 2, pp. 40–41, 1997.doi:10.1145/253671.253695

[10]   S. Khoor, K. Nieberl, K. Fugedi, E. Kail, "Telemedicine ECG-telemetry with Bluetooth technology," *Comput. Cardiol.*, pp. 585–588, 2001.

[11]   E. S. Hall *et al.*, "Enabling remote access to personal electronic medical records," *IEEE Eng. Med. Biol. Mag.*, vol. 22, no. 3, pp. 133–139, May/Jun. 2003.

[12]   J. Andreasson *et al.*, "Remote system for patient monitoring using Bluetooth/spl trade," Sensors, 2002, *Proc. IEEE*, vol. 1, pp. 304–307, Jun. 2002.

[13]   C. Finch, "Mobile computing in healthcare," *Health Manage. Technol.*, vol. 20, no. 3, pp. 63–64, Apr. 1999.

[14]   CHILI (tele-) radiology workstation (Steinbeis Transferzentrum Medizinische Informatik, Heidelberg, Germany). A joint project with the German Cancer Research Center in Heidelberg, Germany.

[15]   U. Engelmann, A. Schroter, E. Boralv, T. Schweitzer, and H.-P. Meinzer, "Mobile teleradiology: All images everywhere," *Int. Congr. Series 1230*, pp. 844–850, 2001. doi:10.1016/S0531-5131(01)00143-1

[16]   R. Andrade, A. von Wangenheim, and M. K. Bortoluzzi, "Wireless and PDA: A novel strategy to access DICOM-compliant medical data on mobile devices," *Int. J. Med. Inf.*, vol. 71, pp. 157–163, 2003.doi:10.1016/S1386-5056(03)00093-5

[17]   I. Maglogiannis, N. Apostolopoulos, and P. Tsoukias, "Designing and implementing an electronic health record for personal digital assistants (PDA's)," *Int. J. Qual. Life Res.*, vol. 2, no. 1, pp. 63–67, 2004.

[18]   V. Stanford, "Pervasive health care applications face tough security challenges," *Pervasive Comput.*, Apr./Jun. 2002.

[19]   F. A. Mora, G. Passariello, G. Carrault and J.-P. Le Pichon "Intelligent patient monitoring and management systems: A review," *IEEE Eng. Med. Biol. Mag.*, vol. 12, no. 4, pp. 23–33, Dec. 1993.

[20] S. Barro, J. Presedo, D. Castro, M. Fernandez-Delgado, S. Fraga, M. Lama, and J. Vila, "Intelligent telemonitoring of critical-care patients," *IEEE Eng. Med. Biol. Mag.*, vol. 18, no. 4, pp. 80–88, Jul./Aug. 1999.

[21] A. Kara, "Protecting privacy in remote-patient monitoring," *Comput. Practices*, May 2001.

[22] J. Dan and J. Luprano, "Homecare: A telemedical application," *Med. Device Technol.*, Dec. 2003.

[23] J. Luprano, "Telemonitoring of physiological parameters: The engineer point-of-view (sensors and data transmission)," *2003 SATW Congr.*, Bern, Sep. 2003.

[24] K. Hung and Y. Ting, "Implementation of a WAP-based telemedicine system for patient monitoring," *IEEE Trans. Inf. Technol. Biomed.*, vol. 7, no. 2, Jun. 2003.

[25] R. Ashok, D. Agrawal, "Next-generation wearable networks," *IEEE Computer*, pp. 31–39, Nov. 2003.

CHAPTER 10

Telemedicine and Virtual Reality

10.1 INTRODUCTION

When one thinks of research fields related to life sciences, virtual reality (VR) may not be the first one that comes to mind, since it is usually related with the entertainment industry. However, the advent of low-cost specialized motion capture devices and powerful graphics rendering hardware have provided computers with advanced visualization capabilities, able to illustrate medical information with a multitude of techniques. Formally, VR can be thought of as a collection of capturing devices and interaction of hardware and software that caters for immersing and navigating a user within a synthetic virtual environment, possibly with novel, naturalistic means of interfacing, while also providing interaction capabilities with objects contained therein; in the case of augmented reality (AR), this notion can be extended with images, video footage, or models of real objects and environments [7, 9].

To fully immerse one into such a synthetic environment, usually takes more than a computer screen and a mouse; instead, users typically wear helmets and gloves, used to provide location-specific renderings of the virtual world in the form of a heads-up display (HUD), as well as provide tactile feedback in case the user interacts with a particular object. These devices also serve as input devices to the computer that renders the virtual environment, since they collect information concerning the users' gaze direction and capture their hand movements and gestures; as a result users, who are represented via an "avatar" within the virtual environment, can communicate their position, direction, and movement to the host computer and express interest or manipulate a model of an object using the same interaction means as they would in the physical world. The result of this naturalistic process is the complete immersion of the user into the virtual environment, where complex or ancient buildings or remote locations are easy to reproduce and offer as a setting. For example, models of ancient cities such as Pompeii or Olympia, museums like the Louvre, or distant planets such as Mars are typical "destinations" for the user of a virtual world. In science-related applications medical practitioners can plan and test difficult operations, robots can be tele-operated in space, and science teachers can demonstrate different forms of gravity in schools. As a result, VR is quickly catching up in related fields, especially where visualization, navigation, and interaction with multidimensional data are required.

10.2   OVERVIEW OF VIRTUAL REALITY TECHNOLOGY

In general, the minimum set of features for VR applications are visualization and navigation of data collected and rendered from a host computer. This information can be completely synthetic, as in the case of architecture, or remotely captured and reconstructed, as is the case with medical applications such as magnetic resonance imaging (MRI) or positron emission tomography (PET) data [1].

The computer-generated environment within which the use agent exists consists of a three-dimensional (3D) reconstruction, or rendering, of a containing environment, filled with objects and possibly other user agents; to create such an environment, one typically needs dedicated modeling software, even if models of real objects and environments are used. In order to achieve greater realism and thus enhance the immersion experience, texture maps can be used to reproduce the visual surface detail of a real object in the models, while real-time simulations of dynamics allow objects to be moved within the virtual environment and interact with each other following natural laws such as gravity and friction.

As mentioned before, the conventional method of working with computers, that is providing information via a mouse or keyboard and viewing results on a monitor, can hamper immersion into a virtual world. Therefore, VR developers tend to replace this interface with one that is more intuitive and allows interaction with everyday, human–human communication concepts; this naturalistic interface typically consists of related hardware containing capture sensors (cameras, IR, magnetic, etc.) for controlling user interaction within the virtual world and feedback effectors (such as headphones, goggles, or force-feedback gloves) for the user [2].

10.3   MEDICAL APPLICATIONS OF VIRTUAL REALITY TECHNOLOGY

In the field of medical applications, VR is usually put to use in surgical and nonsurgical training and planning, medical imaging and visualization, and rehabilitation and behavioural evaluation [9]. In the field of telemedicine, telesurgery and data visualization are the prevalent applications of VR, since its real power is shown in situations where multidimensional, heterogeneous information may be visualized collectively and transparently to the end user. In this framework, it is the medical practitioners who are immersed into the virtual environment and either navigate into 3D data sets such as MRI scans or transfer their knowledge and expertise into remote areas. For example, situations where a patient cannot be moved to a hospital for examinations or surgery or cases where specialists are located very far from the point of interest, e.g., an accident scene can benefit from telesurgery or remote data visualization [7, 8, 10].

Putting VR technologies to work in the field of telesurgery brings in a number of major advantages related to resources, safety, and interoperability. For example, there is a limited need to use corpses or experimental subjects that are typically costly to maintain. In addition to this, new surgical techniques can be tested without putting both patients and practitioners to danger; in this case, surgery progress can be monitored from a variety of viewpoints, even from within internal organs. Finally, all transactions can be recorded and then recreated, so that annotation and discussion in a later stage are possible.

In order to design and deploy usable VR applications in the field of telemedicine, one must take into account the following requirements:

1. *Display fidelity*: High-resolution graphics are essential in order to illustrate reconstructed organs in a truthful manner.

2. *Reconstruction of organs*: Modeled organs should have the same density and texture as the real ones.

3. *Organ reactions and dynamics*: Modeled organs should be in the same positions and have the same functionality as the real ones in simulated circumstances, for example, bleeding.

4. *Organ/object interaction*: In cases of simulated probing or surgery, a dynamics simulation should exist so that the modeled organs respond to pressure or cuts by the virtual instruments.

5. *Force or haptic feedback*: Touching a virtual organ with a virtual instrument should be communicated to the user via a glove or other feedback device [4, 5].

Another important field where VR applications can prove useful is exposure treatment and rehabilitation. In these cases, novel and untested or dangerous techniques can be designed and investigated before large-scale deployment. In exposure treatment, the patient is gradually immersed in situations or subjected to stimuli that may be the cause of phobias or extreme reactions (for example, acrophobia, the fear of height, or disorders in war veterans [11]), usually via mounting a HUD. Monitoring devices that capture a variety of possible factors, e.g., respiration, heart beat, gaze direction, etc. are connected to a host computer to help locate pathological patterns or identify the source of the problem. After this happens, patients may be gradually transferred to *in vivo* situations and exposed to real stimuli [3].

In a rehabilitation scenario, patients may be immersed in entertaining or collaborative VR environments, thus improving their performance in related situation, while still catering for easy monitoring and recording of their performance. In one of the best known practices [6], patients with Parkinson's in San Anselmo, California, wear head-mounted devices and are immersed in an environment where virtual objects are places near their feet; trying to avoid these virtual objects help patients improve their mobility skills, a factor which is severely debilitated by the particular disease. VR treatment, besides being safer and less costly, can assist in reducing related medication; for example, if mobility disorders are reversed via the aforementioned technique, drugs with questionable side effects, like Ldopa, cease to be required for treatment.

10.4   CONCLUSIONS

VR tools and techniques are being developed rapidly in the scientific, engineering, and medical areas. This technology will directly affect medical practice. Computer simulation will allow physicians to practice surgical procedures in a virtual environment in which there is no risk to patients, and where mistakes can be recognized and rectified immediately by the computer. Procedures can be reviewed from new, insightful perspectives that are not possible in the real world.

The innovators in medical VR will be called upon to refine technical efficiency and increase physical and psychological comfort and capability, while keeping an eye on reducing costs for healthcare. The mandate is complex, but like VR technology itself, the possibilities are very exciting. While the possibilities—and the need—for medical VR are immense, approaches and solutions using new VR-based applications require diligent, cooperative efforts among technology developers, medical practitioners, and medical consumers to establish where future requirements and demand will lie.

REFERENCES

[1]   P. Aslan, B. Lee, R. Kuo, R. K. Babayan, L. R. Kavoussi, K. A. Pavlin, and G. M. Preminger, "Secured medical imaging over the Internet," in *Medicine Meets Virtual Reality: Art, Science, Technology: Healthcare Evolution*. San Diego, CA: IOS Press, 1998, pp. 74–78.

[2]   E. A. Attree, B. M. Brooks, F. D. Rose, T. K. Andrews, A. G. Leadbetter, and B. R. Clifford, "Memory processes and virtual environments: I can't remember what was there, but I can remember how I got there. Implications for persons with disabilities," in *Proc. European Conf. Disability, Virtual Reality, and Associated Technology*, 1996, pp. 117–121.

[3]   R. W. Bloom, "Psychiatric therapeutic applications of virtual reality technology (VRT): research prospectus and phenomenological critique," in *Medicine Meets Virtual Reality: Global Healthcare Grid*. San Diego, CA: IOS Press, 1997, pp. 11–16.

[4]   G. Burdea, R. Goratowski, and N. Langrana, "Tactile sensing glove for computerized hand diagnosis," *J. Med. Virtual Real.*, vol. 1, pp. 40–44, 1995.

[5]   M. Burrow, "A telemedicine testbed for developing and evaluating telerobotic tools for rural health care," in *Medicine Meets Virtual Reality II: Interactive Technology and Healthcare: Visionary Applications for Simulation Visualization Robotics*. San Diego, CA: Aligned Management Associates, 1994, pp. 15–18.

[6]   A. Camurri, E. Cervetto, B. Mazzarino, P. Morasso, G. Ornato, F. Priano, C. Re, L. Tabbone, A. Tanzini, R. Trocca, and G. Volpe, "Application of multimedia techniques in the physical rehabilitation of Parkinson's patients," in *Proc. 1st Int. Workshop Virtual Reality Rehabilitation*, Lausanne, Switzerland, 2002, pp. 65–75.

[7]   A. Rovetta, M. Canina, F. Lorini, R. Pegoraro *et al.*, "Demonstration of surgical telerobotics and virtual telepresence by Internet + ISDN from Monterey (USA) to Milan (Italy)," in *Proc. Medicine Meets Virtual Reality: Art, Science, Technology: Healthcare Evolution.* J. D. Westwood, H. M. Hoffman, D. Stredney, S. Weghorst, Eds. San Diego, CA: IOS Press, 1998, pp. 79–83.

[8]   C. V. Edmond, D. Heskamp, D. Sluis, D. Stredney, D. Sessanna, G. Weit, R. Yagel, S. Weghorst, P. Openheimer, J. Miller, M. Levin, and L. Rosenberg, "ENT endoscopic surgical training simulator," in *Medicine Meets Virtual Reality: Global Healthcare Grid.* San Diego, CA: IOS Press, 1997, pp. 518–528.

[9]   W. Greenleaf and T. Piantanida, "Medical applications of virtual reality technology," in *The Biomedical Engineering Handbook*, J. D. Bronzino, Ed., 2nd ed. Boca Raton, FL: CRC Press, 2000, pp. 69.1–69.23.

[10]   J. W. Hill, J. F Jensen, P. S. Green, and A. S. Shah, "Two-handed tele-presence surgery demonstration system," in *Proc. ANS 6th Ann. Topical Meeting Robotics and Remote Systems*, vol. 2, Monterey, CA, 1995, pp. 713–720.

[11]   S. Mrdeza and I. Pandzic, "Analysis of virtual reality contribution to treatment of patients with post-traumatic stress disorder," in *Proc. 7th Int. Conf. Telecommunications CONTel 2003*, Zagreb, Croatia.

[12]   G. Riva, "The emergence of e-health: Using virtual reality and the Internet for providing advanced healthcare services," *Int. J. Healthc. Technol. Manage.*, vol. 4, no. 1–2, pp. 15–40, 2002.doi:10.1504/IJHTM.2002.001126

The Authors

ILIAS G. MAGLOGIANNIS

Ilias G. Maglogiannis was born in Athens, Greece in 1973. He received a Diploma in Electrical and Computer Engineering and a Ph.D. degree in Biomedical Engineering and Medical Informatics from the National Technical University of Athens (NTUA), Greece in 1996 and 2000, respectively, with scholarship from the Greek Government. From 1996 until 2000 he worked as a researcher in the Biomedical Engineering Laboratory in NTUA and he has been engaged in several European and National Projects. From 1998 until 2000 he was also the head of the computer laboratory, in the Department of Civil Engineering in the NTUA. From February 2001 until March 2004 he served as a Visiting Lecturer in the Department of Information and Communication Systems Engineering in University of the Aegean. In March 2004 he was elected Lecturer in the same Department.

His published scientific work includes 5 lecture notes (in Greek) on Biomedical Engineering and Multimedia topics, 21 journal papers, and more than 40 international conference papers. He has served on program and organizing committees of national and international conferences and he is a reviewer for several scientific journals. His scientific activities include biomedical engineering, telemedicine and medical informatics, image processing, and multimedia. Dr. Maglogiannis is a member of IEEE—Societies: Engineering in Medicine and Biology, Computer, Communications, SPIE—International Society for Optical Engineering, ACM, the Technical Chamber of Greece, the Greek Computer Society, and the Hellenic Organization of Biomedical Engineering. Dr. Maglogiannis is also a national representative for Greece in the IFIP Working Group 12.5 (Artificial Intelligence—Knowledge-Oriented Development of Applications).

KOSTAS KARPOUZIS

Kostas Karpouzis is an associate researcher at the Institute of Communication and Computer Systems (ICCS) and holds an adjunct lecturer position at the University of Piraeus, teaching data warehousing and data mining. He graduated from the School of Electrical and Computer Engineering of the National Technical University of Athens in 1998 and received his Ph.D. degree in 2001 from the same University. His current research interests lie in the areas of human computer interaction, image and video processing, image interchange infrastructures using the MPEG-4 and MPEG-21 standards, sign language synthesis, and virtual reality. Dr. Karpouzis has published more than 70 papers in international journals and proceedings of international conferences. He is a member of the technical committee of the International Conference on Image Processing (ICIP), the IFIP Conference on Artificial Intelligence Applications and Innovations (AIAI) and a reviewer in many international journals. Since 1995 he has participated in more than 12 R&D projects at Greek and European level. He is also a national representative in IFIP Working Groups 12.5 "Artificial Intelligence Applications" and 3.2 "Informatics and ICT in Higher Education."

MANOLIS WALLACE

Manolis Wallace was born in Athens in 1977. He obtained his degree in Electrical and Computer Engineering from the National Technical University of Athens (NTUA) and his Ph.D. from the Computer Science Division of NTUA. He is with the University of Indianapolis, Athens Campus (UIA) since 2001 where he serves now as an Assistant Professor. Since 2004 he is also the Chair of the Department of Computer Science of UIA.

His main research interests include handling of uncertainty, information systems, data mining, personalization, and applications of technology in education. He has published more than 40 papers in the above-mentioned fields, 10 of which are in international journals. He is the guest editor of 2 journal special issues and a reviewer for more than 10 journals and various books. His academic volunteering work also includes participation in various conferences as organizing committee chair, session organizer, session chair, or program committee member.

Printed in the United States
by Baker & Taylor Publisher Services